# SAFETY TRAINING METHODS

# SAFETY TRAINING METHODS
## Practical Solutions for the Next Millennium

Second Edition

JACK B. ReVELLE, PhD
JOE STEPHENSON, PhD, CSP

A Wiley-Interscience Publication
**JOHN WILEY & SONS, INC.**
New York / Chichester / Brisbane / Toronto / Singapore

This text is printed on acid-free paper.



***Library of Congress Cataloging in Publication Data:***

Re Velle, Jack B.
Safety training methods : practical solutions for the next millennium / by Jack B. Re Velle and Francis J. Stephenson — 2nd ed.
p. cm.
"A Wiley-Interscience publication."
Includes bibliographical references and index.
ISBN 0-471-55230-5 (acid-free paper)
1. Safety education, Industrial. I. Stephenson, Francis J. II. Title.
HD7260.6.R48 1995 363.11'7'07—dc20 94-18137

Printed in the United States of America

10 9 8 7 6 5 4 3 2 1

# DEDICATION

This revised edition of *Safety Training Methods* is dedicated to all the members of the safety and health community, but especially to those who sought information and guidance in the first edition. Our special thanks are offered to those health providers whose professional competence and caring were instrumental to restoring wellness to Joe. We extend a very special thanks to our families and friends whose love, support, and patience encouraged us to persevere. Without these important people, this edition would not be possible.

# CONTENTS

# PREFACE

In 1980 when *Safety Training Methods* was initially developed for publication, I had already been a practicing safety professional for over a decade. About 10 years later, when the time came to prepare a revised edition, I had been too long away from daily involvement in the practice of safety management, safety engineering, and safety training to produce a worthy successor to the original edition. For me to independently prepare a really quality manuscript would have required an extraordinary amount of time and effort for research since the safety profession has progressed so far since my original effort.

Therefore, I made the very difficult decision to share authorship so that the revised edition, this book, could be produced as effectively and efficiently as possible. After a nationwide search that lasted over two years, I heard about Joe Stephenson. After we spoke on the telephone, Joe provided me with impressive evidence of his capability for research and capacity for clear written communication. I knew my search was complete and urged him to join me as co-author of *Safety Training Methods.*

What follows my introductory remarks are Joe's. I believe you will find them to be as motivating as I did.

Jack B. Re Velle, PhD

*Consultant to Industry*
*Orange, California*

Nothing affects the quality of life more than your health.

In June of 1992 I was told that I had cancer that had spread to my liver and probably had only a few months to live. Fortunately, I got a second opinion and, after a couple of operations, six weeks of radiation, and a year of chemotherapy, am apparently cancer free and feeling great.

Having been blessed with extremely good health for the first 50 years of my life, my encounter with cancer was a significant learning experience. Prior to my second operation, I thought pain was something like a bee sting.

As I lay there in the hospital whining about my discomfort and inconvenience, I thought about how it must feel to have abdominal muscles penetrated not by a skilled surgeon's scalpel while under general anesthesia, but by a piece of rusty rebar or angle iron while wide awake . . . and as I thought about the effect of my situation on my family, I thought about the effect on the families of the victims of occupational injuries and illnesses and realized, as never before, the importance of our work as safety and health professionals.

If my opening sentence is indeed true, then protecting health and safety should, logically, be an extremely high priority for each of us as individuals and for our businesses, our nation, and for society.

Education and training have always been essential elements in protecting safety and health. In fact, I would argue that it may be the most important of the three "Es" of safety (Education, Engineering, and Enforcement). While totally convinced that good engineering and enforcement are equally essential (all are legs on a three-legged stool), both good engineering and enforcement are largely dependent on the education and training effort.

Good engineering relies on the knowledge and skill of the engineers, managers, end users, and others involved in the design, construction, testing, and maintaining processes. This knowledge and this skill are obtained primarily through education and training.

The most effective enforcement is obtained when all members of the organization know, understand, and appreciate the reasons for and value of safety requirements. In a truly safe work environment, all members of the organization demand, not tolerate, safety rules and regulations. This type "enforcement" is also gained through effective education and training.

I deeply appreciate having had the opportunity to assist Jack Re Velle in updating and upgrading his highly successful first edition of *Safety Training Methods.*

We both hope that the information contained in *Safety Training Methods* will aid those attempting to protect the safety and health of the workforce in accomplishing this very noble task.

JOE STEPHENSON, PhD, CSP

*West Coast Regional Manager*
*COBRO Corporation*

# SAFETY TRAINING METHODS

## CHAPTER 1

# FUNDAMENTALS OF SAFETY AND HEALTH TRAINING

### INTRODUCTION

Before developing, implementing, evaluating, or upgrading any environment, safety, and health training program, it is important to have a clear understanding of the fundamentals of what, when, where, why, how, by whom, and to whom safety and health training should be presented.

### WHAT

The "whats" of any training program are (or should be) developed initially through the application of job and task analysis. All organizations should continually analyze jobs and tasks to determine their training needs.

This is particularly important as changes are introduced into the workplace. A detailed training needs analysis should be performed every time a new job, task, process, procedure, facility, or piece of equipment is introduced into the workplace.

However, for many jobs and activities, the training needs analysis has already been accomplished by regulatory agencies, and the training requirements are outlined or specified by codes, standards, or regulations.

This is particularly true of safety and health training requirements, many of which are specified by the Occupational Safety and Health Act in 29 CFR 1910 (General Industry Standards) and/or 29 CFR 1926 (Construction Industry Standards). See Chapter 3.

A comprehensive safety and health education and training program includes all employees, beginning the first day on the job and continuing throughout their employment.

In today's rapidly changing environment, it is more important than ever that employee education and training, including safety and health education and training, be a continuously improving, ongoing process. Today's skills and knowledge will not be adequate to protect us from tomorrow's hazards. A comprehensive safety and health training program will include timely orientations and training for all levels of the organization and methods of ensuring that safety and health-related skills and knowledge continue to be developed and upgraded throughout the employee's career with the organization (Figure 1.1).

## General Employee Orientation Programs

Initial safety and health training and orientation should include the following information:

- The rights and responsibilities of employees under the OSHAct, including conduct on the part of employees that may be necessary to comply with OSHA standard.

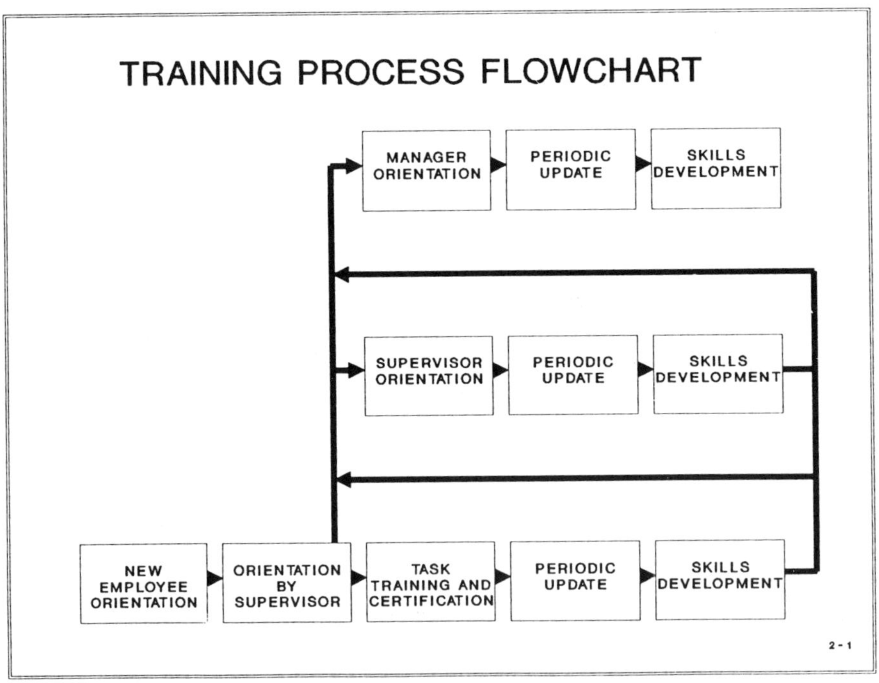


**FIGURE 1.1** Training process flowchart.

- First aid and emergency procedures.
- The need to report all events resulting in injury, illness, or property damage, as well as near misses.
- The need to report hazardous conditions and practices.
- Company and departmental safety and health rules and their application to employees' job.
- Specific hazards of the employees' job and how to avoid them.
- Procedures, precautions, safeguards, and personal protective equipment necessary to protect employees from the hazards of the job.

Safety and health training for a new employee includes the four Rs: responsibilities, rights, rules, and regulations:

- Teaching a new employee his or her *responsibilities* under the OSHAct.
- Teaching a new employee his or her *rights* under the OSHAct.
- Teaching a new employee the general and departmental safety and health *rules* of the company.
- Teaching a new employee starting a job new to him or her about the applicable OSHAct safety and health *regulations* (standards).

Safety and health training should also include information about action to be taken in the following emergency situations:

- What to do if the employee or a fellow employee is seriously injured or develops severe symptoms of illness.
- How and when to evacuate the building rapidly.
- How and when to give a fire alarm to fellow employees, report a fire to the local fire department, and to use portable fire extinguishers and other available fire fighting equipment.
- How and when to shut off electricity, gas, and other special hazards.

It is especially important for employees to know what to do in emergency situations because quick action is required.

Since the employer is responsible for correcting hazards and recording illnesses and injuries caused by occupational hazards, a comprehensive training program will teach employees to report such conditions and events. New employees should be told about the reporting requirements of the OSHAct. Since regulations promulgated under the OSHAct require employers to maintain specific information on all recordable injuries and illnesses, the need for employees to report all job-related injuries and illnesses that occur in their workplaces should be stressed.

It is a good idea to have employees report all injuries and illnesses so that management can decide whether an event is recordable. Then too, some "near misses" that are not actually recordable have come very close to being serious enough to count. Something can often be learned from a near miss that will help prevent the real event from happening.

Safety and health training must start the first day on the job. An orientation should be held at each level of the organization, with each orientation reinforcing the previous one and becoming more specific in terms of potential hazards and safety requirements as the employee gets "closer" to his or her workplace. In addition to learning key safety rights and responsibilities by repetition, a safety orientation at each level conveys the message early that "These people are *serious* about safety and really do care about my wellbeing!" (Figure 1.2).

The importance of prompt reporting of hazardous conditions or practices should be explained during safety and health training before an employee starts a new job. In addition, safety and health training for new employees should include general and departmental safety and health rules.

In all but very small companies, new employee orientations should be provided at multiple organizational levels, beginning with a broad company-level orientation covering all key environment, safety, and health information on a general level, and ending with a very specific one-on-one orientation delivered by the employee's direct supervisor (and augmented by fellow workers). In larger companies there may be one or more intermediate orientations. There should be an orientation at each major level (division, department, etc.) of the organization. The information will tend to get more personal and more specific at each level. A multilevel approach ensures that the employee gets "the big picture" by having key information repeated. After hearing the really important safety rights and responsibilities and key elements of the safety program repeated at different levels, the new employee is expected to remember the key elements and develop the attitude that "this company must be serious about safety!"

NEW
EMPLOYEE
ORIENTATION

WHAT: Company safety and health requirements

WHEN: Immediately after being hired/transferred

WHERE: Classroom/Inprocessing Location

WHY: Provide general safety rights and responsibilities

HOW: Lecture, Video, Interactive Video, Computer Based Program augmented with employee handbook/handout materials

BY WHOM: Safety or training staff

TO WHOM: All newly hired or transferred employees

**FIGURE 1.2** New Employee orientation.

Even though the company-level safety orientation and the orientations at successive levels of the organization are important, the most important safety training an employee can receive is the safety orientation given by his or her supervisor the first day on the job. To this point, much of the information has been general. It may have been provided by individuals who will have infrequent contact and little or no direct impact on the new employee. Now the new employee must have this general information made specific and personal by having his or her boss explain, in detail, the specific potential hazards and the safety rules, regulations, and requirements that apply at the actual work location. Additionally, of course, the supervisor is expressing his or her commitment to safety and personalizing his or her expectations and the consequences of failing to meet those expectations. At the conclusion of this orientation the new employee should fully understand what is expected, should have a "warm fuzzy feeling" that the organization and the supervisor are really dedicated to providing a safe workplace, and should also realize that safety requirements are "real!" (Figure 1.3).

Safety orientation of the new employee is the job of everyone from the general manager, to mid-managers, to the training and safety staffs, to supervision, to fellow workers who look after him or her during the first days on the job. The orientation procedure should have two objectives:

- Employee protection: to impart the factual information necessary to the individual's safety.
- Employee involvement: to make him or her an active member of the safety team.

Even though the nature and specificity of the information will change, a common format can be used for all levels of new employee environmental safety and health orientations. The general format of new employee safety and

ORIENTATION
BY
SUPERVISOR

WHAT: Job specific safety and health requirements

WHEN: Within 24 hours of reporting to jobsite (after inprocessing)

WHERE: Jobsite

WHY: Describe job/jobsite specific hazards and safety and health rights and responsibilities

HOW: Lecture/discussion, jobsite tour, handout materials

BY WHOM: First line supervisor

TO WHOM: All newly hired or transferred employees

**FIGURE 1.3** Orientation by supervisor.

health orientations at all levels of the organization can and probably should be the same, with key information being repeated and reinforced and with information being more specific and detailed at each level of the organization (top to bottom). (Figure 1.4).

## Specific New Job and Task Training

Before an employee can be expected to work safely, he or she must be shown the safe way to do the job. Before being exposed to potential job hazards, the ability of each employee to correctly (safely) perform each task should be confirmed. For simple, low hazard tasks this may require only a short, informal discussion with the supervisor. However, for new, complex, or high hazard operations it may require specific training and/or performance testing (Figure 1.5).

A wide variety of training areas should be considered in most companies. The actual job and task training should be based on job and task analysis and on regulatory (OSHA, MSHA) requirements.

This is generally the most complex portion of the environmental safety and health training program to manage. This occurs because of the need for individual training for each class and level of worker, and because job-driven and regulatory-driven requirements tend to change (increase) rapidly.

Even though most of this training will be directed to the new employee, it is also important that this training be conducted for employees who change jobs and for new jobs or tasks introduced into the workplace.

NEW EMPLOYEE ENVIRONMENT, SAFETY & HEALTH ORIENTATION

WORK ACTIVITIES (Describe the work process and surroundings)

POTENTIAL HAZARDS (Describe possible environment, safety & health hazards associated with work activities and the workplace)

STEPS TAKEN TO PROTECT EMPLOYEES (Describe the procedures, barriers, personal protective equipment, etc. provided)

EMPLOYEE RIGHTS AND RESPONSIBILITIES (Describe the safety and health rights of employees and the things that each employee must do to ensure a safe and healthful workplace)

EMERGENCY ACTIONS (Outline emergency procedures to include accident reporting requirements)

QUESTIONS (Provide an opportunity for employees to clarify information provided and to seek new information)

2-4

**FIGURE 1.4** Orientation format.

TASK
TRAINING AND
CERTIFICATION

WHAT: Job and site specific task procedures and requirements

WHEN: Before performing any task for the first time or after significant changes in process, location, or procedures

WHERE: Jobsite and/or classroom

WHY: Provide detailed task procedures and requirements

HOW: Lecture, demonstration, practical exercise, examination

BY WHOM: Supervision and/or safety or training staff

TO WHOM: All employees

**FIGURE 1.5** Task training.

Job and task training will almost always involve a combination of classroom training, practical exercises, and/or on-the-job training. It is important that on-the-job training be planned, executed, documented, and evaluated with the same care as any other education or training.

Minimum requirements for job and task specific training have been developed by OSHA (and MSHA, for the mining industry). OSHA training requirements are contained in Chapter 3.

### Developmental Training

Developmental training refers to education and training efforts directed at ensuring that employees are regularly provided with current information and that continual improvement of safety, quality, and production efforts is made possible by constant upgrading of individual and organizational skills and capabilities.

Development training is essential to career progression for individuals and continued improvement of organizations' capabilities.

Equipment, procedures, locations, requirements, and associated hazards change at an ever increasing rate. It is imperative that workplace changes and the safety implications of those changes be effectively communicated. Provisions for this type of upgrade training must be integrated into the work process. "Tailgate" and periodic safety meetings are common ways of doing this, however, other methods such as "just-in-time training" offer great potential also (Figure 1.6).

Developmental programs include periodic updates, such as safety and staff meetings, which reinforce key learning, increase awareness of key requirements, and provide job specific instructions for current tasks, and new or changed safety or job information.

**PERIODIC UPDATE**

WHAT: New safety and health or job requirement, awareness training

WHEN: Weekly, monthly, or quarterly depending on nature of work

WHERE: Jobsite or classroom

WHY: Provide new information on changes, motivate, increase safety and health awareness

HOW: Lecture, discussion, demonstration, video

BY WHOM: Supervisors, managers, safety/training staff, or any individual member of the organization

TO WHOM: All employees

**FIGURE 1.6** Periodic update.

Developmental programs also include nonmandatory education and training aimed at enhancing skills and capabilities of individuals or organizations. Examples include computer skills training, general supervision, management and communications skills training, and total quality management programs. These efforts may or may not include specific environment, safety, and health applications, but all tend to enhance the overall safety and health effort by upgrading the competency level within the organization.

For a person to advance in responsibility, position, and value within an organization and to respond to changes, that person must develop and enhance existing job skills and knowledge, including safety and health-related skills and knowledge (Figure 1.7).

Developmental programs may also include degree and nondegree programs offered by local trade schools, community or junior colleges, or colleges and universities. These programs may or may not have formal ties to the company and to company education and training efforts. Typically, support for these programs is limited to tuition assistance and/or scheduling concessions to make participation possible.

## Supervisor/Management Training

In addition to the orientations, job/task training, periodic updates, and skill development programs for general employees, it is also important that parallel training is provided for supervisors and managers at all levels.

The supervisor has long been recognized as a "key person" in any organization. New supervisors must receive the orientation and training required to fully understand their new safety and health responsibilities and training and/or testing must be provided to ensure that supervisors have the skills and knowledge required to meet their safety and health responsibilities (Figure 1.8).

SKILLS DEVELOPMENT

WHAT: Supervisory skills, first aid/CPR, computer training, etc. college or trade school courses

WHEN: As needed, before promotion to next level, ongoing effort

WHERE: Classroom, jobsite, off site

WHY: Improve productivity, upgrade skills, prepare for promotion

HOW: Lecture, self-study courses, video, computer based training

BY WHOM: Safety/training staff, vendors, college or trade schools

TO WHOM: All employees

**FIGURE 1.7** Skills development.

Orientation training for new supervisors should be directed at providing a clear understanding of job requirements and the basic knowledge needed to meet those requirements. For example, safety and health responsibilities could include performing worksite inspections, conducting safety meetings and job-specific training, responding to emergency situations, and investigating accidents. Thus the new supervisor may need additional hazard recognition training, a "train the trainer" course, additional emergency response training, and accident investigation training. This training must include the specific administrative requirements associated with each of these tasks, such as how to fill out and distribute required forms and other documentation.

As individuals move "up" the organization, their potential to do "damage" and to do "good" increases dramatically. Management support and commit-

SUPERVISOR ORIENTATION

WHAT: Supervisory safety and health requirements and skills

WHEN: Immediately before or after being hired, promoted, or transferred

WHERE: Classroom

WHY: Provide skills and knowledge to inspect workplace, train subordinates, investigate accidents, and other requirements

HOW: Lecture, Practical exercise, examination, video

BY WHOM: Safety/training staff, management

TO WHOM: All newly hired, promoted, or transferred supervisors

**FIGURE 1.8** Supervisor orientation.

ment are essential to the success of any safety and health effort. Managers must receive the training needed to understand safety and health requirements and to ensure that safety is fully integrated into all organizational activities (Figure 1.9).

Programs for new managers should also concentrate on providing specific information about new requirements and new knowledge or skills required to meet these requirements. This "training" may be in the form of small briefings conducted by the next higher level of management, by members of the executive staff, and by managers of other organizational units.

Periodic update and skill development training for supervisors and managers should parallel the programs for general employees, but be tailored to specific needs. Frequently, most of the skill development training for managers is provided by outside vendors and by participation in executive seminars and conferences.

## WHEN

Safety training should be provided upon employment and thereafter, as needed. Prior to the assumption of specific employment responsibilities, orientation and on-the-job training should be scheduled. Once an employee has been assigned, however, training may be scheduled before, during, or after regularly scheduled working hours. Factors that must be carefully considered in the scheduling of safety training include union agreements, employee preferences, and operational needs as defined by managers and supervisors.

Other timing considerations include the assignment of new or additional responsibilities. It is unfair to both the employee and the company to assign other responsibilities to an employee before the appropriate safety training the been provided.

MANAGER
ORIENTATION

WHAT: Manager safety and health requirements, operations data

WHEN: Immediately before or after being hired, promoted, or transferred

WHERE: Manager's office, conference room

WHY: Provide specific knowledge required to meet safety and health requirements and provide indepth knowledge of organization and operations

HOW: Management briefings, discussions, self-study

BY WHOM: Top management, executive staff, safety and health manager, other managers

TO WHOM: All newly hired, promoted, or transferred managers

**FIGURE 1.9** Manager orientation.

To the extent possible, training programs should be scheduled in conjunction with supervisors' requirements to minimize operational interruptions.

It is important to know when to provide safety and health training. Such material is especially directed to employees performing new jobs to be sure that they are made aware of hazards and maintain this awareness.

The purpose of occupational safety and health training is to make employees aware of the safety and health hazards under which they operate during the working day, as well as to show them how to perform their jobs without endangering themselves or their fellow employees. The following quotations from the Occupational Safety and Health Act of 1970 (OSHAct) standards are examples of the emphasis placed on safety and health training by the Act:

> CFR 1910.178(l) Powered Industrial Trucks. Only trained and authorized operators shall be permitted to operate a powered industrial truck. Methods shall be devised to train operators in the safe operation of powered industrial trucks.

> CFR 1910.217(f)(1) Mechanical Power Presses. The employer shall train and instruct the operator in the safe method of work before starting work on any operation covered by this section.

Before any employee starts a job that is new to him or her, it is the company's responsibility to provide safety and health training for that specific job. Thus safety and health training is required for a new employee, an employee who is changing jobs, or an employee who is about to start a newly created job or whose job has been changed. The immediate supervisor is the best person to provide safety and health training for employees starting jobs new to them, because this individual knows the most about the hazards involved in the jobs under his or her supervision.

Training an employee before work is started on a job affords a good opportunity to influence the formation of good work habits. With newly hired employees who are beginning their first jobs, appropriate training can help in the formation of safe and healthful work habits that should continue throughout their entire working lives. With newly hired employees who have past work experience, an employer can get them off on the right foot. With experienced employees who have been assigned to new jobs or to newly created jobs, or to changed jobs within the organization, appropriate training can detect and correct unsafe or unhealthful work habits of which the employees may not even be aware. Providing safety and health training for employees starting new jobs or jobs new to them is an important supervisory function.

As previously stated, employee safety and health education and training must begin immediately upon hire and must continue throughout employment. It is imperative that employees receive key safety and health training *before* being exposed to job hazards.

## WHERE

When planning safety training programs, their locations must be considered. Depending on the type of training to be scheduled, a broad variety of locations should be considered:

- Workplace
- Simulated workplace (trainer)
- Office
- Classroom
- Laboratory
- Outside company premises.

Precisely which location(s) are selected for use depends on the type of training, the availability of certain locations, the willingness of managers and supervisors to release operational sites otherwise allocated to production activities, the suitability of the locations to the presentation of training, and similar factors.

## WHY

The reasons for providing safety and health training are basically the same as for providing a safety and health program:

- **It is Good Business.** Employees who know how to do the right things right the first time and are motivated to work correctly (safely) are efficient and productive. They also prevent and avoid costly accidents and mistakes.
- **It is the "Right Thing to Do."** Employers generally are not in business "just for the money" (otherwise everyone would sell illegal drugs or rob banks). Most employers take pride in producing a product or service that benefits the community or society. Most people have sincere feelings about the welfare of the people with whom they work. Nothing is more demoralizing for all concerned than a serious workplace accident, especially one in which someone is killed or seriously wounded. This is especially true if the accident was preventable (and virtually all are).
- **It is the Law.** Employers have a legal obligation to provide training required by OSHA or other regulatory agencies. Employers also have a legal obligation to provide a safe and healthful workplace. This requires training of the workforce. Penalties can be severe for failure to comply. Penalties may include OSHA or other fines and litigation, which may include civil suits and criminal prosecution.

In addition to providing key skills and knowledge, training can be used to motivate and to modify behavior and attitudes.

Effective training is a cornerstone and an essential element of any environmental safety and health program.

## HOW

The Occupational Safety and Health Administration (Pamphlet #12, *Safety and Health Training* of the Safety and Health Assistance Series written by the Office of Consultation Services, September 1991 draft) offers the following advice on how to design, conduct, evaluate, and document safety and health training:

### Design

#### *Pretraining Elements*

Before training begins, be sure that your company policy clearly states the company's commitment to health and safety, including the training program. This commitment must include paid work time for training and training in the language that the worker understands. Involve both management and employees in the development of the training. Be sure that health and safety concerns can be improved by training, and that specific goals for the training are clearly identified.

#### *Teaching and Learning Principles*

Training your supervisors and employees need not be complex or lengthy. In most small businesses, the owner or manager can do the training. In larger businesses which have extensive training needs, more formal training may be necessary. In either case, five basic principles should guide your training program.

1. **Perceived Purpose.** The learner must understand the purpose for the learning. Therefore, the beginning of any training program should focus on why this instruction will be useful to the learner.
2. **Order of Presentation.** Training should be presented so that the organization and meaning are clear to the learners. For example, if you are teaching employees the proper way to use a respirator, the order in which you present the material should match the steps the employee must use to choose, fit, wear, and maintain the respirator.
3. **Appropriate Practice.** We learn best when we can immediately practice and apply newly acquired knowledge and skills. Therefore, instruction for job safety and health is best given at the worksite where demonstration, practice and application can be immediate. When onsite instruction is not feasible, plan for practice and application of the new knowledge and skills as soon as possible.

4. **Knowledge of Results.** As we practice, we need to know as soon as possible if we are correct. Incorrect practice can establish harmful patterns while praise for correct actions will enhance motivation as well as begin to establish the desired results.
5. **Individual Differentiation.** Since we are individuals, we learn in different ways. A successful training program incorporates a variety of learning opportunities, such as written instruction, audiovisual instruction, lectures and coaching in hands-on situations. Also, we learn at different speeds, and we learn through teaching others. Therefore, after the initial instruction and some practice, dividing the group into teaching and learning pairs, pairing a slower learner with a quicker one is often helpful.

### *Developing Learning Activities*

Keeping in mind the Five Learning Principles described above, develop your learning activities to meet the training needs you have identified. Be imaginative in your choice of methods and materials. Make full use of your resources. You can get ideas about training by looking at similar training programs offered by other companies in the same industry, or by obtaining suggestions from such organizations as the National Safety Council, the American Society of Safety Engineers, the Bureau of Labor Statistics, OSHA consultation programs, and the OSHA Office of Training and Education.

### *Conducting Training*

Employees must be motivated enough to want to learn and to improve. You can enhance the success of the training by:

1. Preparing employees for training by putting them at ease.
2. Recruiting employees who would be good trainers of their peers and preparing them to do training.
3. Explaining the job or training topic and determining how much your employees already know about it.
4. Boosting their interest in learning about the job by helping them see how the training will help them.
5. Pacing the instruction to the speed of the learners, presenting it clearly, and being patient.
6. Presenting no more information in one session than your employees can master.
7. Having your employees perform each step of the operation and repeat your instructions as to why it is done the way you describe. Have them repeat a sequence until you are satisfied that they know how to do it.
8. Encouraging employees to help each other by dividing the group into teacher/learner pairs or practice pairs.
9. During the initial practice period, check on the employees frequently for correct performance. Taper off on the checking as they become more proficient.
10. Encouraging your employees to incorporate the new skill into the way they work best, but caution them not to change the newly learned procedure unless they first check with you or their supervisor.

### Evaluation

A plan for evaluating the training session(s) should be generated as the needs are being identified and the content is being developed, not delayed until the training is completed. Evaluation will help employers and supervisors ensure that the desired learning has been achieved and that the employee's performance has improved on the job. Among the methods of evaluating training are:

1. Asking employees' opinions, using questionnaires or during informal discussions, to assess the relevance and appropriateness of the training program.
2. Testing the results of the training by stopping at the work station to observe an employee's performance both before and after training, noting improvement or changes. After the training, you can measure its success a second time by asking the worker for an explanation of the hazards involved in the work and how the worker can protect himself/herself. If the training has been well-presented and understood, each trained worker should be able to give a brief but comprehensive response and to show performance improvement or change.
3. Recording workplace improvements such as reduced injury or accident rates.

However it is conducted, an evaluation of training can provide employers the information necessary to decide whether or not their employees achieved the desired results, whether the training session should be offered again at some future date, and if any modifications need to be made in the training program.

### Recordkeeping

Even if you operate a very small business, it is to your advantage to keep training records. A simple format is all that is needed: who was trained, on what topic or job, for how long, on what date, and an evaluation of the learner's participation and success. These records will help you make sure that everyone who needs training will receive it, that refresher courses will be provided at regular intervals, and that documentation will be available, if necessary, to show that training was done appropriately. (An example of a training record form found in Appendix 1 of OSHA Pamphlet #12 is included in Chapter 6, "Safety Training Records.")

### Sources for Assistance

In addition to the pamphlets in this series, assistance with training programs or the identification of resources for training is available through such organizations as

- OSHA full-service Area Offices
- State agencies which have their own OSHA-approved occupational safety and health programs
- OSHA-funded State onsite consultation programs for employers, local safety councils

- the OSHA Office of Training and Education and OSHA-funded New Directions grantees.

Specific training requirements for OSHA standards and guidelines can be found in OSHA publication 2254 (Current edition).

Every instructor, whether neophyte or old timer, has specific responsibilities associated with his or her presentation of safety and health training programs. These are as follows:

- **Size of Class.** A class should be limited to 15 to 25 students. If the number in the group to be instructed exceeds 25, more than one class should be formed.
- **Meeting Place.** The room should be well-ventilated, well-lighted, and away from noise and other distractions. The room should be clean, orderly, and attractive.
- **Equipment.** The meeting place should be equipped with a whiteboard and dry erase pens (or equivalent), erasers, and a pointer. It should have tables or other writing space for each student. Seats should be comfortable and arranged so that all the students can easily see and hear. A "U"-shaped seating arrangement, with the instructor at the open end of the "U," is sometimes ideal. Auditorium seating may also be used. The room should be provided with a screen and the appropriate projector(s) for showing overhead transparencies, motion pictures, slides, or other visual aids. A video monitor or television set and video cassette recorder (VCR) should also be available. If computer training or a computer demonstration is involved, equipment to project the computer screen may also be required.
- **The Instructor's Job.** The most important assets of an instructor are sincerity, enthusiasm, and a knowledge of the subject. It is not necessary to have a teaching or training background, but good presentation skills are needed. A sincere desire to help others and enthusiasm about teaching are basic to effective instruction. The procedures outlined in this book will help the instructor present the material clearly and concisely.

Assignments should be given in advance so students have an opportunity to read the material and can then discuss the contents during the class period. However, this may not be possible for the beginning session.

In training program discussions, the leader encourages students to think and to talk. The best tools for promoting discussion are specific topics and provocative questions. For this reason, prepared meeting plans include a number of topics and questions. After a discussion is started, the leader guides it with additional questions, but not those that can be answered "yes" or "no." Participation should be obtained from the entire group. One person should not be allowed to do all the talking.

Skill in leading a discussion is acquired with practice. Every time an instructor leads a discussion, he or she has an opportunity to improve.

- **Use of Visual Aids.** It is highly recommended that visual aids such as video cassettes, overhead transparencies, or slides be previewed before showing to the class. In addition, aids such as personal protective equipment, guards, charts, and models may be used to good advantage.
- **Use of Tests.** Tests help the instructor emphasize and reinforce the key points of a lesson. They also help the instructor in determining how good a job he or she is doing and provide excellent program documentation. Tests not only provide objective proof that an individual attended the class, they also document the level of understanding of key points. Both pretests and posttests should be considered. *It may sometimes be desirable to call tests "written reviews" since some people tend to be intimidated by the term "test."* In many instances, the term "written review" is actually more descriptive of the actual intent, which should be to reinforce and evaluate understanding. The "test" may also be oral or involve actual performance of certain tasks. Some formal "qualification" or "certification" programs may require a combination of written and performance evaluation.
- **Use of Discussion Topics.** Discussion topics should be listed for each chapter emphasized in the class discussion. Since there may be others that apply to a specific operation or industry, the discussions should not be restricted to the topics listed here. It is a good idea to write on the whiteboard the points agreed on by the class. This will help to remind everyone of what has been discussed and to make sure that no important point is overlooked. The instructor is in the best position to determine the points that need stressing.
- **Lesson Plans.** Effective teaching calls for good planning. "Lesson plans," which should be written, standardize training. They also help the instructor to accomplish the following:
  - Present material in the proper order
  - Avoid omission of essential material
  - Conduct the sessions according to a timetable
  - Place proper emphasis on items to be covered
  - Provide for student participation
  - Gain confidence

For lesson plans to be really effective, each instructor should carefully plan the presentation. See Figure 1.10 for a sample lesson plan format.

As previously discussed, upon completion of a training program, an examination (announced at the beginning of the program) should be administered and the grade recorded. Documentation of participation should be maintained by name, dates, total hours, grade on final examination, and so on to satisfy review by OSHA compliance officers.

**LESSON PLAN**

TITLE: ______________________ LP NUMBER: ________

LENGTH: ______________ DATE PREPARED/REVISED: ____________

REFERENCES:

TRAINING AIDS:

LEARNING OBJECTIVES:

| OUTLINE | TIME | AUDIO/VISUAL AID |
|---|---|---|
| | | |

**FIGURE 1.10** Lesson plan format.

In addition, it may be possible to evaluate the quality of company training programs and personnel by observing correlations between examination scores and accident frequency and severity rates before and after training by department and by shift. This is discussed further in Chapter 13.

At the end of each course, attendees should also be asked to complete a course critique. The course critique is an important evaluation tool and is also discussed in detail in Chapter 13.

## WHO

Just who provides instruction for specific topics depends largely on the subjects to be taught, who is available, who is qualified, who has the ability to make effective presentations, and so on. With these factors considered, choose from among the following persons to provide various type of safety training programs.

- Director of training
- Training department personnel
- Director of safety
- Plant engineer
- Safety engineer
- Industrial hygienist
- Industrial engineer
- Environmental engineer
- First-line supervisors
- Consultants
- Insurance company safety specialists
- Fire department personnel
- Medical personnel
- Machine, vehicle, and equipment operators
- Peers

The nature and complexity of the program will determine the mix of organizational units and individuals who actually present the instruction.

Regulatory agencies and internal requirements will determine the prerequisites for instructor qualification. All persons delivering training should have some "train-the-trainer" instruction and some oversight or other methods of quality assurance. This may range from very formal programs with rigid instructor certification and program accreditation requirements to a simple review of course critiques.

To ensure that high quality safety and health training is provided to all company personnel, the safety manager and other members of the management team should review the proposed course outline and materials. In larger organizations, a Training Advisory Group may be formed. This is particularly recommended in organizations where environmental safety and health training is provided by a centralized training organization.

The function of the Training Advisory Group is to review (approve) courses, instructors, and overall training management of safety and health training programs and to provide technical assistance to the training organization. This group should include safety and health professionals and management and craft representatives from the client organizations.

The level of authority of the advisory group may vary. The group should almost always have review responsibilities. The level of its approval authority will be determined by the composition of the Training Advisory Group, the level of expertise available in the training organization, and the organization being held accountable for meeting compliance requirements. For example, if there are extremely well-qualified safety and health professionals assigned to the training department, it probably would not be necessary or appropriate for the S&H Training Advisory Group to have approval authority over lesson plans and course materials as long as course objectives meet regulatory requirements.

## TO WHOM

In general terms, all members of the organization should receive basic safety and health training.

In addition, OSHA pamphlet #12 recommends the following:

> **Some Usual Types of Specialized Training**
>
> *Safety and Health Training for Managers.* A good safety and health program is impossible without support and understanding from the top. Training for managers in their responsibilities is necessary to ensure their continuing support and understanding. Formal classroom training may not be necessary. The subject can be covered periodically as a part of regular management meetings.
>
> Managers should understand the importance of the goal and objectives set for safety and health and clearly communicate them to their employees. They should remember to make clear assignments of safety and health responsibilities, provide authority and resources to carry them out and hold subordinate managers and supervisors accountable. It is important that they let employees see their support for the safety and health program and that they set a good example by scrupulously following all the safety and health rules. They should also actively encourage employee involvement in safety and health problem identification and resolution. (For further information on management roles, see Pamphlet #3 of this series, "Visible Management Leadership: Showing Your Management Commitment.")
>
> These topics can be covered and illustrated with examples in a relatively short time. They should be repeated at least annually.
>
> *Safety and Health Training for Supervisors.* Although all employees should be involved in matters of safety and health, often workers are promoted to supervisory positions without adequate knowledge of how to train other employees in the safe and proper way to do the job and without full knowledge of the company's policies and procedures. They may need additional training in detecting and controlling hazards, investigating accidents, their role in ensuring the maintenance of physical protections, handling emergencies and, in general, how to supervise.

Since supervisors do a lot of on-the-job training, they will also need training in how to train and how to reinforce training. They will also need help in learning how to apply fair and consistent discipline, when necessary. Such training can be provided by the supervisor's immediate management, by the Safety Department or by outside sources such as the OSHA Consultation Service. The OSHA Training Institute is one good source for some of the training.

## SUMMARY

OSHA Pamphlet #12 summarizes safety and health training as follows:

> Good safety and health training programs will benefit the employee, the employer, and the company as a whole because such programs demonstrate an honest concern for health and safety. They help improve workers' morale and often decrease near-miss incidents and accidents, resulting in fewer lost work days and more production. Records of such training provide evidence that employees have been trained and can be used to determine what training employees still need to take.
>
> Safety and health training programs consist of determining training needs, involving employees in the program design and in the teaching. New hires, contract workers and workers in high risk areas should be especially targeted for training. Use the five principles of teaching and learning when constructing course content. Evaluate each course and modify when necessary. Always keep orderly records, noting not only who was trained, but also that the trainee learned what was taught.

This chapter has provided a broad general overview of the fundamentals of safety and health training. The balance of this book will provide more detailed information and expand on these fundamentals. For example, Chapter 3 and most of Chapters 6–12 are largely devoted to expanding upon "what" should be taught. Chapter 2 details the roles and responsibilities of those involved in the safety and health training effort and Chapters 4, 5, and 13 discuss the "hows" of presenting, documenting, and evaluating safety and health training.

# CHAPTER 2

# ROLES AND RESPONSIBILITIES IN SAFETY TRAINING

## INTRODUCTION

The purpose of this chapter is to provide guidance in determining and clearly documenting the roles and responsibilities of various members of the organization in the overall safety training effort.

Obviously, the size and nature of the organization will significantly influence specific responsibilities. For example, in a small organization without a separate training department, the development and presentation of instruction may be the responsibility of the safety department or, in a very small organization, the safety professional or even the supervisor. In larger organizations, a training department, with a staff of instructional designers, training analysts, and instructors, may have the responsibility for developing and presenting safety training and safety and health professionals and supervisors may be used as subject matter experts only to aid in the process.

To present information in a way that is meaningful to all readers, roles and responsibilities will be outlined for various specialties that may be found in larger organizations. In smaller organizations, many of these roles may be consolidated.

## DEFINITIONS

**Manager/Supervisor:** An individual in the organizational chain of command with direct oversight and control of one or more subordinate employees.

**Employee:** Any member of the organization.

**First Line Supervisor:** The last supervisor in the chain of command. The direct supervisor of individuals who do not have subordinates.

**Safety and Health Professional:** An occupational safety professional, industrial hygienist, health physicist, occupational medicine professional, environmental specialist, or other individual with professional qualifications and specific safety and/or health-related duties.

**Safety and Health Manager:** An individual heading a safety and health organization. Actual job title may be safety or safety and health director, administrator, or coordinator; Chief Loss Prevention Engineer; etc.

**Training Manager:** An individual heading a training organization. Actual job title may be training director, administrator, coordinator, etc.

**Instructional Designer:** An individual qualified by education, training, and/or experience to design and develop instructional programs. Specifically skilled in performing needs analysis, design, development, implementation, and evaluation tasks in the training program development process.

**Training Analyst:** An individual who assists managers, supervisors, subject matter experts, instructional designers, and others involved in the training program development process in analyzing training needs and requirements to determine the most cost-effective methods of presentation and training alternatives.

**Instructor:** An individual who conducts training presentations.

**Subject Matter Expert:** An individual who, based on education, training, and/or experience has developed indepth knowledge and mastery of a specific subject. Typical subject matter experts for safety-related training programs include safety and health professionals who have expertise in specific safety and health disciplines and supervisors who have indepth knowledge of specific job tasks and processes.

**Quality Assurance Manager:** An individual heading an organization responsible for quality assurance.

**Procurement Manager:** An individual heading a procurement organization or operation.

**Vendor:** An outside individual or organization providing training services on a contract or fee basis.

## RESPONSIBILITIES

### Managers/Supervisors

Managers and supervisors at all levels have overall responsibility for the safety training of their subordinates. This includes determining safety and health training needs, providing adequate training resources, and enforcing training requirements.

## Employees

All members of the organization are responsible for ensuring that they do not proceed with any job without having the knowledge and skills to perform the job safely and correctly. All employees must inform their supervisors in cases where their training or job knowledge is inadequate. Additionally, all employees are responsible for attending assigned training, actively participating in class activities, and providing feedback to the training organization and/or supervision/management as to the effectiveness of the training. Finally, all employees are responsible for applying the information provided in training courses to their specific jobs, i.e. for doing things the way they were trained to do them.

## First Line Supervisor

Even though every member of management from plant manager to first-time supervisor is directly responsible for the safety of the employees under him or her, the first-time supervisor is the key to a good safety program, including the safety training effort.

In addition to the responsibilities listed above for all managers and supervisors, the first line supervisor must also integrate safety training into job instructions and, because he or she is directly observed by the workforce, must also lead by example.

A supervisor sets the pace and greatly influences the employee's interest in safety by his or her actions. The examples set by the supervisor should reflect sincerity and alertness. Some specific suggestions are given below.

- Observe all safety and fire protection rules. The supervisor who makes exceptions to such rules for personal convenience seriously undermines the safety effort in that area.
- Demonstrate sincere concern about safety by seriously and consistently "preaching and practicing" safety.
- Keep safety on at least an equal level of importance with quality and quantity of output.
- Ignore no violation of safety regulations.
- Wear personal protective equipment where required. It is a good way to "sell" it and to demonstrate that using such protection is the intelligent thing to do.
- Discuss some aspect of safety with employees every day.
- Be enthusiastic about safety and fire protection. The enthusiasm that the supervisor displays will generate enthusiasm in the employees. Give safety priority among your problems. Never let quality, production, or cost considerations compromise safety or essential fire protection.

"Teaching takes place most effectively when people see that they are successful—when they experience a "series of successes." The first line supervisor should suggest ways of putting the "S.O.S." principle to work in safety training. Suggestions could include the following:

Make safety a part of job instruction.

Instruct not only the new employee in safe practices, but also the older employee who has been transferred to a new job.

Draft understandable safety rules.

Make sure instructions are easy to follow.

Always give reasons for safety instruction.

Have employees "feed back" instructions to you, so you know they know.

Give recognition for what is done correctly.

To provide accurate and relevant safety training, the first line supervisor must appreciate and evaluate fully the safety and fire hazards involved on a given job and must understand thoroughly the entire process or operation for which he or she is responsible.

The ability of an employee to perform a specific job is dependent on his or her education, training, experience, and general capabilities. To achieve the safest, most efficient performance, the supervisor must know these characteristics when planning job assignments, training programs, and performance reviews.

An effective accident prevention program requires proper job performance from everyone in the workplace. The first line supervisor has a major role in influencing job performance by providing job-specific training.

All employees must know about the materials and equipment they are working with, what hazards are known in the operation, and how these hazards have been controlled or eliminated.

Each individual employee needs to know and understand the following points (especially if they have been included in the company policy and in a "code of safe practices"):

No employee is expected to undertake a job until he or she has received instruction on how to do it properly and has been authorized to perform that job.

No employee should undertake a job that appears to be unsafe.

Mechanical safeguards are in place and must be kept in place.

Each employee is expected to report all unsafe conditions encountered during work.

Even slight injury or illness suffered by an employee must be reported at once.

The safety and health program can be amplified as needs and new methods are discovered.

In addition to the points above, any safety rules that are a condition of employment, such as the use of safety shoes or eye protection, should be explained clearly and enforced at once.

Some specific training requirements in the OSHA standards must be met, such as those that pertain to first aid and powered industrial trucks (including forklifts). In general, they deal with situations where the use of untrained or improperly trained operators on skill machinery could cause hazardous situations to develop, not only for the operator, but possibly for nearby workers, too.

Particular attention must be given to new employees. Immediately on arriving at work, new employees begin to learn things and to form attitudes about the company, the job, their boss, and their fellow employees. Learning and attitude formation occur regardless of whether the employer makes a training effort. If the new employees are trained during those first few hours and days to do things the right way, considerable losses may be avoided later.

At the same time, attention must be paid to regular employees, including the old timers. Old habits can be wrong habits. An employee who continues to repeat an unsafe procedure is not working safely, even if an "accident" has not resulted from this behavior.

Although every employee's attitude should be one of determination that "accidents" can be prevented, one thing more may be needed. It should be stressed that the responsibility assigned to the person in charge of the job—as well as to all other supervisors—is to be sure that there is a concerted effort underway at all times to follow every safe work procedure and health practice applicable to that job. It should be clearly explained to these supervisors that they should never silently condone unsafe or unhealthful activity in or around any workplace.

Finally, here are some less specific signals that might indicate a need for training or retraining:

Excessive waste or scrap

High labor turnover

An increase in the number of near misses that could have resulted in accidents

A recent upswing in actual "accident" experience. High injury and illness incidence. Expansion of business and/or new employment.

A change in process or a new process

The first line supervisor is responsible for recognizing and providing for (directly or through company training programs) the safety training needs of his or her organizational unit.

### Safety and Health Professionals

Safety and health professionals are responsible for identifying situations, conditions, operations, and processes where training is an appropriate hazard control and where safety and health training is needed. They are also responsible for providing oversight of the safety and health training program to ensure that OSHA and other training requirements are being met. Safety and health professionals should serve as subject matter experts to assist in developing safety and health training programs and frequently will also serve as safety and health instructors.

### Safety and Health Managers

Safety and health managers are responsible for ensuring that situations, conditions, operations, and processes requiring safety training to control hazards are identified and that appropriate safety and health training programs are developed and implemented. This may require providing subject matter experts, instructors, and other resources to the training organization.

Safety and health managers are also responsible for ensuring that OSHA and other mandated safety and health training requirements are met.

### Training Managers

Training managers are responsible for ensuring that safety and health training courses required by line managers and by the safety and health manager are developed (or procured) and presented.

### Instructional Designers

Instructional designers are responsible for translating the training needs identified by managers into specific training courses or programs. This includes conducting (with or without the aid of training analysts) job, task, and needs analyses, designing and developing learning objectives, lesson plans, training aids, student handout materials, practical exercises, tests, and other course materials, and producing implementation and evaluation plans.

### Training Analysts

Training analysts are responsible for coordinating the efforts of managers, supervisors, subject matter experts, instructional designers, instructors, and others involved in the training program development or procurement process. The training analyst serves as a single point of contact and coordinates/

manages the program development process. Training analysts also aid in determining which training programs should be developed inhouse and which training programs should be purchased or obtained from vendors.

### Instructors

Instructors have the primary responsibility for presenting training. This includes reviewing and updating lesson plans and course materials, conducting lectures and practical exercises, presenting audiovisual programs, administering and grading tests, and collecting, evaluating, and responding to course critiques.

### Subject Matter Experts

Subject matter experts are responsible for providing technical advise in the preparation or procurement of training courses or programs. This includes assisting instructional designers, trainers, and instructors by describing specific job tasks, operations, and/or processes. It also includes reviewing all course materials for technical accuracy, aiding in developing specification for vendor-provided training, and evaluating vendor qualifications and proposals.

### Quality Assurance Managers

Quality assurance managers are responsible for ensuring that safety and health training is conducted and procured in accordance with established quality program standards.

### Procurement Managers

Procurement managers are responsible for ensuring that training materials and services meet the requirements determined by line managers, training analysts, and subject matter experts.

### Vendor

Vendors are responsible for providing training services as specified in contracts or other agreements.

As previously stated, in smaller organizations, many of these functions may be combined. For example, a safety and health professional may serve as subject matter expert, instructional designer, training analyst, and instructor. You will need to group, consolidate, or modify the above responsibilities to fit your organization.

SAFETY & HEALTH
TRAINING RESPONSIBILITIES

MANAGERS/SUPERVISORS
Identify requirements and provide training resources

EMPLOYEES
Participate in training and apply information

FIRST LINE SUPERVISORS
Conduct job specific instructions. Enforce requirements

SAFETY AND HEALTH PROFESSIONALS
Identify requirement. Serve as subject matter experts.

SAFETY AND HEALTH MANAGERS
Identify requirements. Provide subject matter experts.

TRAINING MANAGERS
Ensure safety & health courses are developed and presented

INSTRUCTIONAL DESIGNERS
Translate training needs into training courses

TRAINING ANALYSTS
Coordinate training program development or procurement

INSTRUCTORS
Present training. Review/update lesson plans & materials

SUBJECT MATTER EXPERTS
Provide technical advice. Review and approve courses

QUALITY ASSURANCE MANAGERS
Provide program oversight

PROCUREMENT MANAGERS
Ensure procured services and materials meet specifications

VENDORS
Meet contract requirements

4-1

**FIGURE 2.1** Safety training responsibilities.

## SUMMARY

The responsibilities for safety training are spread throughout the organization, with generally everyone responsible for aiding in identifying training needs, supervisors and managers responsible for ensuring that employees are trained, and the various members of the safety and training communities responsible for ensuring that technically correct training is cost-effectively developed or procured, presented, evaluated, and documented (Figure 2.1).

Safety and health training responsibilities are obviously only one part of the overall safety and health responsibilities of an organization, but a very important part.

# CHAPTER 3

# OSHA TRAINING REQUIREMENTS

## INTRODUCTION

In addition to the safety and health training that an organization develops and implements to aid employees in identifying and avoiding the workplace hazards that may be specific to a given location or operation, the Occupational Safety and Health Act of 1970, as outlined in Title 29 of the Code of Federal Regulations (29 CFR), mandates certain training requirements.

Parts of 29 CFR containing specific training requirements include:

General Industry Training Requirements, Part 1910
Maritime Training Requirements, Parts 1915, 1917, 1918
Construction Training Requirements, Part 1926
Agriculture Training Requirements, Part 1928
Federal Employee Training Requirements, Part 1960

Because training requirements are spread throughout the Occupational Safety and Health Act, it is difficult to identify and extract all of the many training requirements.

To aid in this task, the Occupational Safety and Health Administration produces a document entitled "Training Requirements in OSHA Standards and Training Guidelines." The original version was produced in 1979. The current version (as of October 1992) was published in 1992. The document number is OSHA 2254 (Revised) and is available free (single copies) from the OSHA Publications Office, U.S. Department of Labor, 200 Constitution Avenue, Room N 3101, N.W., Washington, DC 20210.

The following pages were extracted from OSHA 2254 (Revised) and include the General Industry training requirements from 29 CFR 1910 and the Construction Industry training requirements from 29 CFR 1926.

Relatively new OSHA requirements include new or amended requirements for hazard communications, tag and lockout, and bloodborne pathogens training. Additional discussion of these requirements is included in Chapter 9, "Hazard Communications," Chapter 10, "Tag and Lockout," and Chapter 11, "First Aid and CPR," respectively.

## GENERAL INDUSTRY TRAINING REQUIREMENTS

The following training requirements have been excerpted from Title 29, Code of Federal Regulations Part 1910. Note that additional training requirements may appear in certain other standards (ANSI, NFPA, etc.) adopted by reference in Part 1910 and therefore mandatory.

| Subject and Standard Number | Training Requirement |
|---|---|
| EMPLOYEE EMERGENCY PLANS AND FIRE PREVENTION PLANS<br>1910.38(a)(5)(i),(ii), and (iii) | (i) Before implementing the emergency action plan, the (iii) employer shall designate and train a sufficient number of persons to assist in the safe and orderly emergency evacuation of employees.<br>(ii) The employer shall review the plan with each employee covered by the plan at the following times:<br>(a) Initially when the plan is developed,<br>(b) Whenever the employee's responsibilities or designated actions under the plan change, and<br>(c) Whenever the plan is changed.<br>(iii) The employer shall review with each employee upon initial assignment those parts of the plan which the employee must know to protect the employee in the event of an emergency. The written plan shall be kept at the workplace and made available for employee review. For those employers with 10 or fewer employees the plan may be communicated orally to employees and the employer need not maintain a written plan. |
| 1910.38(b)(4)(i) and (ii) | (i) The employer shall apprise employees of the fire hazards of the materials and processes to which they are exposed.<br>(ii) The employer shall review with each employee upon initial assignment those |

parts of the fire prevention plan which the employee must know to protect the employee in the event of an emergency. The written plan shall be kept in the workplace and made available for employee review. For those employers with 10 or fewer employees, the plan may be communicated orally to employees and the employer need not maintain a written plan.

OPERATIONS-TRAINING
1910.66(i)(ii through iv)

(i) Working platforms shall be operated only by persons who are proficient in the operation, safe use and inspection of the particular working platform to be operated.

(ii) All employees who operate working platforms shall be trained in the following:

(A) Recognition of, and preventive measures for, the safety hazards associated with their individual work tasks.

(B) General recognition and prevention of safety hazards associated with the use of working platforms, including the provisions in the section relating to the particular platform to be operated.

(C) Emergency action plan procedures required in paragraph (E)(9) of this section.

(D) Work procedures required in paragraph (i)(1)(iv) of this section.

(E) Personal fall arrest system inspection, care, use and system performance.

(iii) Training of employees in the operation and inspection of working platforms shall be done by a competent person.

(iv) Written work procedures for the operation, safe use and inspection of working platforms shall be provided for employee training. Pictorial methods of instruction, may be used, in lieu of written work procedures, in employee communication is improved using this method. The operating manuals supplied by manufacturers for platform system components can serve as the basis for these procedures.

(v) The employer shall certify that employees have been trained in operating and inspecting a working platform by preparing a certification record which includes the identity of the person trained, the signature of the person trained, the signature of the employer or the person who conducted the training and the date that training was completed. The certification record shall be prepared at the completion of the training required in paragraph (i)(1)(ii) of this section, and shall be maintained in a file for the duration of the employee's employment. The certification record shall be kept readily available for review by the Assistant Secretary of Labor or the Assistant Secretary's representative.

MANLIFTS
1910.68(e)(i)

(1) *Frequency.* All manlifts shall be inspected by a competent designated person at intervals of not more than 30 days. Limit switches shall be checked weekly. Manlifts found to be unsafe shall not be operated until properly repaired.

VENTILATION
1910.94(d)(9)(i) and (vi)

(i) All employees working in and around open surface tank operations must be instructed as to the hazards of their respective jobs, and in the personal protection and first aid procedures applicable to these hazards.

(vi) . . . Respirators shall be approved by the U.S. Bureau of Mines, U.S. Department of the Interior [see 30 CFR Part II] and shall be selected by a competent industrial hygienist or other technically qualified source. Respirators shall be used in accordance with 51910.134, and persons who may require them shall be trained in their use.

1910.94(d)(ii)(v) and (vi)

(v) If, in emergencies, such as rescue work, it is necessary to enter a tank which may contain a hazardous atmosphere, suitable respirators, such as self-contained breathing apparatus; hose mask with blower, if there is a possibility of oxygen deficiency; or a gas mask, selected and operated in accordance with paragraph (d) (9) (vi) of this section, shall be used. If a containment in the tank can cause dermatitis, or be absorbed through the skin, the employee entering the tank shall also wear protective clothing. At least one trained standby employee, with suitable respirator, shall be present in the nearest uncontaminated area. The standby employee must be able to communicate with the employee in the tank and be able to haul him out of the tank with a lifeline if necessary.

(vi) Maintenance work requiring welding or open flame, where toxic metal fumes such as cadmium, chromium, or lead may be evolved, shall be done only with sufficient local exhaust ventilation to prevent the creation of a health hazard, or be done with respirators selected and used in accordance with subparagraph (9) (vi) of this paragraph. Welding, or the use of open flames near any solvent cleaning equipment shall be permitted only after such equipment has first been thoroughly cleared of solvents and vapors.

OCCUPATIONAL NOISE EXPOSURE
1910.95(k) and (i)

(k) *Training program.*

(1) The employer shall institute a training program for all employees who are exposed to noise at or above an 8-hour time-weighted average of 85 decibels, and shall ensure employee participation in such program.

(2) The training program shall be repeated annually for each employee included in the hearing conservation program. Information provided in the training program shall be updated to be consistent with changes in protective equipment and work processes.

(3) The employer shall ensure that each employee is informed of the following:

(i) The effects of noise on hearing;

(ii) The purpose of hearing protectors, the advantages, disadvantages, and attenuation of various types, and instructions on selection, fitting, use, and care; and

(iii) The purpose of audiometric testing, and an explanation of the test procedures.

(l) Access to information and training materials.

(1) The employer shall make available to affected employees or their representatives copies of this standard and shall also post a copy in the workplace.

(2) The employer shall provide to affected employees any informational materials pertaining to the standard that are supplied to the employer by the Assistant Secretary.

(3) The employer shall provide, upon request, all materials related to the employer's training and education program pertaining to this standard to the Assistant Secretary and the Director.

IONIZING RADIATION
1910.96(f)(3)(viii)

(viii) All employees whose work may necessitate their presence in an area covered by the signal shall be made familiar with the actual sound of the signal—preferably as it sounds at their work location. Before placing the system into operation, all employees normally working in the area shall be made acquainted with the signal by actual demonstration at their work locations.

1910.96(i)(2)

(2) All individuals working in or frequenting any portion of a radiation area shall be informed of the occurrence of radioactive materials or of radiation in such portions of the radiation area; shall be instructed in the safety problems associated with exposure to such materials or radiation and in precautions or devices to minimize exposure; shall be instructed in the applicable provisions of this section for the protection of employees from exposure to radiation or radioactive materials; and shall be advised of reports of radiation exposure which employees may request pursuant b the regulations in this section.

HYDROGEN
1910.103(c)(4)(ii)

(ii) *Attendant.* A qualified person shall be in attendance at all times while the mobile hydrogen supply unit is being unloaded.

FLAMMABLE AND COMBUSTIBLE LIQUIDS
1910.106(b)(S)(vi)(v)(2) and (3)

(2) That detailed printed instructions of what to do in flood emergencies are properly posted.

(3) That station operators and other employees depended upon to carry out such instructions are thoroughly informed as to the location and operation of such valves and other equipment necessary to effect these requirements.

EXPLOSIVES AND BLASTING AGENTS
1910.109(d)(1)(iii)

(iii) Explosives shall not be transferred from one vehicle to another within the confines of any jurisdiction (city, county, State, or other area) without informing the fire and police departments thereof. In the event of breakdown or collision the local fire and police departments shall be promptly notified to help safeguard such emergencies. Explosives shall be transferred from the disabled vehicle to another only when proper and qualified supervision is provided.

1910.109(d)(2)(iii)(b) (b) Extinguishers shall be filled and ready for immediate use and located near the driver's seat. Extinguishers shall be examined periodically by a competent person.

1910.109(d)(3)(i) (ii) Vehicles transporting explosives shall only be driven by and be in the charge of a driver who is familiar with the traffic regulations, State laws, and the provisions of this section.

1910.109(d)(3)(iii) (iii) Every motor vehicle transporting any quantity of Class A or Class B explosives shall, at all times, be attended by a driver or other attendant of the motor carrier. This attendant shall have been made aware of the class of the explosive material in the vehicle and of its inherent dangers, and shall have been instructed in the measures and procedures to be followed in order to protect the public from those dangers. He shall have been made familiar with the vehicle he is assigned, and shall be trained, supplied with the necessary means, and authorized to move the vehicle when required.

1910.109(9)(g)(iii)(a) (iii) (a) The operator shall be trained in the safe operation of the vehicle together with its mixing, conveying, and related equipment. The employer shall assure that the operator is familiar with the commodities being delivered and the general procedure for handling emergency situations.

1910.109(9)(g)(3)(iv)(c) (c) A qualified person shall evaluate all systems to determine if they will adequately dissipate static under potential field conditions.

1910.109(9)(g)(3)(iv)(c)
1910.109(g)(5)(vii) (vii) Every warehouse used for the storage of blasting agents shall be under the supervision of a competent person.

1910.109(9)(g)(6)(ii) (ii) Vehicles transporting blasting agents shall only be driven by and be in charge of a driver in possession of a valid motor vehicle operator's license. Such a person shall also be familiar with the States vehicle and traffic laws.

1910.109(h)(4)(i)(a)

(a) Vehicles used over public highways for the bulk transportation of water gels or of ingredients classified as dangerous commodities shall meet the requirements of the Department of Transportation and shall meet the requirements of paragraphs (d) and (g)(6) of this section.

STORAGE AND HANDLING OF LIQUEFIED PETROLEUM GASES
1910.110(b)(16)

(16) *Instructions.* Personnel performing installation, removal, operation, and maintenance work shall be properly trained in such function.

1910.110(d)(12)(i)

(i) When standard watch service is provided, it shall be extended to the LP-Gas installation and personnel properly trained.

1910.110(h)(11)(vii)

(vii) The dispensing of LP-Gas into the fuel container of a vehicle shall be performed by a competent attendant who shall remain at the LP-Gas dispenser during the entire transfer operation.

STORAGE AND HANDLING OF ANHYDROUS AMMONIA
1910.111(b)(1)(iv)

(iv) It is a custom-designed and custom-built unit, which no nationally recognized testing laboratory, or Federal, State, municipal or local authority responsible for the enforcement of a Federal, State, municipal, or local law, code or regulation pertaining to the storage, transportation and use of anhydrous ammonia is willing to undertake to accept, certify, list, label or determine to be safe, and the employer has on file a document attesting to its safe condition following the conduct of appropriate tests. The document shall be signed by a registered-professional engineer or other person having special training or experience sufficient to permit him to form an opinion as to safety of the unit involved. The document shall set forth the test bases, test data and results, and also the qualifications of the certifying person.

1910.111(b)(10)(3)(i)

(i) Personnel required to handle ammonia should be trained in safe operating practices and in the proper action to take in the event of emergencies.

1910.111(b)(13)(ii)

(ii) The employer shall insure that unloading operations are performed by reliable persons properly instructed and given the authority to monitor careful compliance with all applicable procedures.

HAZARDOUS WASTE OPERATIONS INTERIM RULE
1910.120(e)

(e) *Training.*

(1) All employees (such as equipment operators and general laborers) exposed to

hazardous substances, health hazards, or safety hazards shall be thoroughly trained in the following:

(i) Names of personnel and alternates responsible for site safety and health;

(ii) Safety, health and other hazards present on the site;

(iii) Use of PPE;

(iv) Work practices by which the employee can minimize risks from herds;

(v) Safe use of engineering controls and equipment on the site;

(vi) Medical surveillance requirements including recognition of symptoms and signs which might indicate over exposure to hazards; and

(vii) Paragraphs (G) through (K) of the site safety and health plan set forth in paragraph (i)(2)(i) of this section.

(2) All employees shall at the time of job assignment receive a minimum of 40 hours of initial instruction off the site, and a minimum of three days of actual field experience under the direct supervision of a trained, experienced supervisor. Workers who may be exposed to unique or special herds shall be provided additional training. The level of training provided shall be consistent with the employee's job function and responsibilities.

(3) On-site management and supervisors directly responsible for or who supervise employees engaged in hazardous waste operations shall receive training as provided in paragraph (e)(1) and (e)(2) of this section and at least eight additional hours of specialized training on managing such operations at the time of job assignment.

(4) Trainer shall have received a level of training higher than and including the subject matter of the level of instruction that they are providing.

(5) Employees shall not participate in field activities until they have been trained to a level required by their job function and responsibility.

(6) Employees and supervisors that have received and successfully completed the training and field experience specified in paragraphs (e)(1), (e)(2), and (e)(3) of this

section shall be certified by their instructor as having completed the necessary training. Any person who has not been so certified or meets the requirements of paragraph (e)(1) of this section shall be prohibited from engaging in hazardous waste operations after March 16,1987.

(7) Employees who are responsible for responding to hazardous emergency situations that may expose them to hazardous substances shall be trained in how to respond to expected emergencies.

(8) Employees specified in paragraph (e)(1) and managers specified in paragraph (e)(3) of this section shall receive eight hours of refresher training annually on the items specified in paragraph (e)(1) of this section and other relevant topics.

(9) Employers who can show by an employee's work experience and/or training that the employee has had initial training equivalent to that training required in paragraphs (e)(1), (e)(2), and (e)(3) of this section shall be considered as meeting the initial training requirements of those paragraphs. Equivalent training includes the training that existing employees might have already received from actual, on-site experience.

1910.120(i)(3)(i)

(3) Off-site emergency response—(i) *Training.* Training for handling emergency responses involving hazardous substances shall be conducted on a monthly basis and shall be at least 24 hours annually. The training shall include as a minimum recognition of hazards, selection, care, and use of personal protective equipment and safe operating procedures to be used at the incident scene.

RESPIRATORY PROTECTION

1910.134(a)(3)

(3) The employee shall use the respiratory protection in accordance with instructions provided and training received.

1910.134(b)(3)

(3) The user shall be instructed and trained in the proper use of respirators and their limitations.

1910.134(e)(2),(3), (4), and (5)(i)

(2) The correct respirator shall be specified for each job. The respirator type is usually specified in the work procedures by a qualified individual supervising the respiratory protective program. The individual

issuing them shall be adequately instructed to insure that the correct respirator is issued. Each respirator permanently assigned to an individual should be durably marked to indicate to whom it was assigned. This mark shall not affect the respirator performance in any way. The date of issuance should be recorded.

(3) Written procedures shall be prepared covering safe use of respirators in dangerous atmospheres that might be encountered in normal operations or in emergencies. Personnel shall be familiar with these procedures and the available respirators.

(4) Respiratory protection is no better than the respirator in use, even though it is worn conscientiously. Frequent random inspections shall be conducted by a qualified individual to assure that respirators are properly selected, used, cleaned, and maintained.

(5) For safe use of any respirator, it is essential that the user be properly instructed in its selection, use, and maintenance. Both supervisors and workers shall be so instructed by competent persons. Training shall provide the men an opportunity to handle the respirator, have it fitted properly, test its face-piece-to-face seal, wear it in normal air for a long familiarity period, and, finally, to wear it in a test atmosphere.

OPERATIONS-TRAINING
1910.66(i)(ii through iv)

(i) Every respirator wearer shall receive fitting instructions including demonstrations and practice in how the respirator should be worn, how to adjust it, and how to determine if it fits properly. Respirators shall not be worn when conditions prevent a good face seal. Such conditions may be a growth of beard, sideburns, a skull cap that projects under the facepiece, or temple pieces on glasses. Also, the absence of one or both dentures can seriously affect the fit of a facepiece. The worker's diligence in observing these factors shall be evaluated by periodic check. To assure proper protection, the facepiece fit shall be checked by the wearer each time he puts on the respirator. This may be done by following the manufacturer's facepiece fitting instructions.

TEMPORARY LABOR CAMPS
1910.142(k)(1) and (2)

(1) Adequate first-aid facilities approved by a health authority shall be maintained and made available in every labor camp for the emergency treatment of injured persons.

(2) Such facilities shall be in charge of a person trained to administer first aid and shall be readily accessible for use at all times.

SPECIFICATIONS FOR ACCIDENT PREVENTION SIGNS AND TAGS
1910.145(c)(1)(ii), (2)(ii) and (3)

(1)(ii) All employees shall be instructed that danger signs indicate immediate danger and that special precautions are necessary.

(2)(ii) All employees shall be instructed that caution signs indicate a possible hazard against which proper precautions should be taken.

(3) *Safety instruction signs.* Safety instruction signs shall be used where there is a need for general instructions and suggestions relative to safety measures.

MEDICAL SERVICES AND FIRST AID
1910.151(a) and (b)

(a) The employer shall insure the ready availability of medical personnel for advice and consultation on matters of plant health.

(b) In the absence of an infirmary, clinic, or hospital in near proximity to the workplace which is used for the treatment of all injured employees, a person or persons shall be adequately trained to render first aid. First aid supplies approved by the consulting physician shall be readily available.

FIRE BRIGADES
1910.156(c)

(c) *Training and education.*

(1) The employer shall provide training and education for all fire brigade members commensurate with those duties and functions that fire brigade members are expected to perform. Such training and education shall be provided to fire brigade members before they perform fire brigade emergency activities. Fire brigade leaders and training instructors shall be provided with training and education which is more comprehensive than that provided to the general membership of the fire brigade.

(2) The employer shall assure that training and education is conducted frequently enough to assure that each member of the fire brigade is able to perform the member's assigned duties and functions satisfactorily and in a safe manner so as not to endanger

fire brigade members or other employees. All fire brigade members shall be provided with training at least annually. In addition, fire brigade members who are expected to perform interior structural fire fighting shall be provided with an education session or training at least quarterly.

(3) The quality of the training and education program for fire brigade members shall be similar to those conducted by such fire training schools as the Maryland Fire and Rescue Institute; Iowa Fire Service Extension; West Virginia Fire Service Exten sion; Georgia Fire Academy, New York State Department, Fire Prevention and Control; Louisiana State University Firemen Training Program, or Washington State's Fire Service Training Commission for Vocational Education. (For example, for the oil refinery industry, with its unique hazards, the training and education program for those fire brigade members shall be similar to those conducted by Texas A & M University, Lamar University, Reno Fire School, or the Delaware State Fire School.)

(3) The employer shall inform fire brigade members about special hazards such as storage and use of flammable liquids and gases, toxic chemicals, radioactive sources, and water reactive substances, to which they may be exposed during fire and other emergencies. The fire brigade members shall also be advised of any changes that occur in relation to the special hazards. The employer shall develop and make available for inspection by fire brigade members, written procedures that describe the actions to be taken in situations involving the special hazards and shall include these in the training and education program.

PARAGRAPH 5 OF APPENDIX A
TO SUBPART L
FIRE PROTECTION
1910.156

5. *Training and education.* The paragraph on training and education does not contain specific training and education requirements because the type, amount, and frequency of training and education will be as varied as are the purposes for which fire brigades are organized. However, the paragraph does require that training and education be commensurate with those functions

that the fire brigade is expected to perform; i.e., those functions specified in the organizational statement. Such a performance requirement provides the necessary flexibility to design a training program which meets the needs of individual fire brigades.

At a minimum, hands-on training is required to be conducted annually for all fire brigade members.

However, for those fire brigade members who are expected to perform interior structural fire fighting, some type of training or education session must be provided at least quarterly.

In addition to the required hands-on training, it is strongly recommended that fire brigade members receive other types of training and education such as: classroom instruction, review of emergency action procedures, prefire planning, review of special hazards in the workplace, and practice in the use of self contained breathing apparatus.

It is not necessary for the employer to duplicate the same training or education that a fire brigade member receives as a member of a community volunteer fire department, rescue squad, or similar organization. However, such training or education must have been provided to the fire brigade member within the past year and it must be documented that the fire brigade member has received the training or education. For example: there is no need for a fire brigade member to receive another training class in the use of positive-pressure self contained breathing apparatus if the fire brigade member has recently completed such training as a member of a community fire department. Instead, the fire brigade member should receive training or education covering other important equipment or duties of the fire brigade as they relate to the workplace hazards, facilities and processes.

It is generally recognized that the effectiveness of fire brigade training and education depends upon the expertise of those providing the training and education as well as the motivation of the fire brigade members. Fire brigade training instructors must receive a higher level of training and education than the fire brigade members they will be teaching. This includes being more knowledgeable about the functions to be performed by the fire brigade and the hazards involved. The instructors should be qualified to train fire brigade members and demonstrate skills in communication, methods of teaching, and motivation. It is important for instructors and fire brigade members alike to be motivated toward the goal of the fire brigade and be aware of the importance of the service that they are providing for the protection of other employees and the workplace.

It is suggested that publications from the International Fire Service Training Association, the National Fire Protection Association (NFPA-1041), the International Society of Fire Service Instructors and other fire training sources be consulted for recommended qualifications of fire brigade training instructors.

In order to be effective, fire brigades must have competent leadership and supervision. It is important for those who supervise the fire brigade during emergency situations, e.g., fire brigade chiefs, leaders, etc., to receive the necessary training and education for supervising fire brigade activities during these hazardous and stressful situations. These fire brigade members with leadership responsibilities should demonstrate skills in strategy and tactics, fire suppression and prevention techniques, leadership principles, pre-fire planning, and safety practices. It is again suggested that fire service training sources be consulted for determining the kinds of training and education which are necessary for those with fire brigade leadership responsibilities.

It is further suggested that fire brigade leaders and fire brigade instructors receive more formalized training and education on a continuing basis by attending classes provided by such training sources as universities and university fire extension services.

The following recommendations should not be considered to be all of the necessary elements of a complete comprehensive training program, but the information may be helpful as a guide in developing a fire brigade training program.

All fire brigade members should be familiar with exit facilities and their location, emergency escape routes for handidicapped workers, and the workplace "emergency action plan."

In addition, fire brigade members who are expected to control and extinguish fires in the incipient stage should, at a minimum, be trained in the use of fire extinguishers, standpipes, and other fire equipment they are assigned to use. They should also be aware of first aid medical procedures and procedures for dealing with special hazards to which they may be exposed. Training and education should include both classroom instruction and actual operation of the equipment under simulated emergency conditions. Hands-on type training must be conducted at least annually but some functions should be reviewed more often.

In addition to the above training, fire brigade members who are expected to perform emergency rescue and interior structural fire fighting should, at a minimum, be familiar with the proper techniques in rescue and fire suppression procedures. Training and education should include fire protection courses, classroom training, simulated fire situations including "wet drills" and, when feasible, extinguishment of actual mock fires. Frequency of training or education must be at least quarterly, but some drills or classroom training should be conducted as often as monthly or even weekly to maintain the proficiency of fire brigade members.

There are many excellent sources of training and education that the employer may want to use in developing a training program for the workplace fire brigade. These sources include publications, seminars, and courses offered by universities. There are also excellent fire school courses by such facilities as Texas A & M University, Delaware State Fire School, Lamar University, and Reno Fire School, that deal with those unique hazards which may be encountered by fire brigades in the oil and chemical industry. These schools, and others, also offer excellent training courses which would be beneficial to fire brigades in other types of industries. These courses should be a continuing part of the training program, and employers are strongly encouraged to take advantage of these excellent resources.

It is also important that fire brigade members be informed about special hazards to which they may be exposed during fire and other emergencies. Such hazards as storage and use areas of flammable liquids and gases, toxic chemicals, water. reactive substances, etc., can pose difficult problems. There must be written procedures developed that describe the actions to be taken in situations involving special hazards. Fire brigade members must be trained in handling these special hazards as well as keeping abreast of any changes that occur in relation to these special hazards.

PORTABLE FIRE EXTINGUISHERS
1910.157(g)

(g) *Training and education.*

(1) Where the employer has provided portable fire extinguishers for employee use in the workplace, the employer shall also provide an educational program to familiarize employees with the general principles of fire extinguisher use and the hazards involved with incipient stage fire fighting.

(2) The employer shall provide the education required in paragraph (g)(3) of this section upon initial employment and at least annually thereafter.

(3) The employer shall provide employees who have been designated to use fire fighting equipment as part of an emergency action plan with training in the use of the appropriate equipment.

(4) The employer shall provide the training required in paragraph (g)(3) of this section upon initial assignment to the designated group of employees and at least annually thereafter.

FIRE EXTINGUISHING SYSTEMS
1910.160(b)(10)

(10) The employer shall train employees designated to inspect, maintain, operate, or repair fixed extinguishing systems and annually review their training to keep them up-to-date in the functions they are to perform.

FIRE DETECTION SYSTEMS
1910.164(c)(4)

(4) The employer shall assure that the servicing, maintenance and testing of fire detection systems, including cleaning and necessary sensitivity adjustments, are performed by a trained person knowledgeable in the operations and functions of the system.

SAFETY RELIEF DEVICES FOR CARGO AND PORTABLE TANKS STORING COMPRESSED GASES
1910.168(b)(4)(ii)

(ii) Only qualified personnel shall be allowed to service safety relief devices. Any servicing or repairs which require resetting of safety relief valves shall be done only by or after consultation with the valve manufacturer.

SERVICING OF SINGLE PIECE AND MULTI-PIECE RIM WHEELS
1910.177(c)(1)(2) and (3). (Includes single piece wheels per Federal Register of multi-piece rim wheels February 3, 1984
(pp. 4338–4352) but not automobile or truck tires marked "LT.")

(c) *Employee training*

(1) The employer shall provide a training program to train and instruct all employees who service multi-piece rim wheels in the hazards involved in servicing and the safety procedures to be followed.

(i) The employer shall assure that no employee services any multi-piece rim wheel unless the employee has been trained and instructed in correct procedures of mounting, truck tires marked "LT," demounting, and all related services, activities, and correct safety precautions for the rim type being serviced, and the safe operating procedures described in paragraph (f) of this section.

(ii) Information to be used in the training program shall include at a minimum, the data contained on the charts and the contents of this standard.

(iii) Where an employer knows or has reason to believe that any of his employees is unable to read and understand the charts or rim manual, the employer shall assure that the employee is instructed concerning the contents of the charts and rim manual in a manner which the employee is able to understand.

(2) The employer shall assure that each employee demonstrates and maintains his ability to service multi-piece rims wheels

safely, including performance of the following tasks:

(i) Demounting of tires (including deflation);

(ii) Inspection of wheel components;

(iii) Mounting of tires (including inflation within a restraining device);

(iv) Use of the restraining device;

(v) Handling of wheels;

(vi) Inflation of tires when a wheel is mounted on the vehicle; and

(vii) Installation and removal of wheels.

(3) The employer shall evaluate each employee's ability to perform these tasks and to service multi-piece rim wheels safely and shall provide additional training as necessary to assure that each employee maintains his proficiency.

POWERED INDUSTRIAL TRUCKS
1910.178(i)

(1) *Operator training.* Only trained and authorized operators shall be permitted to operate a powered industrial truck. Methods shall be devised to train operators in the safe operation of powered industrial trucks.

OVERHEAD AND GANTRY CRANES
1910.179(b)(8)

(8) *Designated personnel.* Only designated personnel shall be permitted to operate a crane covered by this section.

1910.179(o)(3)

(3) *Fire extinguishers.* The employer shall insure that operators are familiar with the operation and care of fire extinguishers provided.

CRAWLER LOCOMOTIVE AND TRUCK CRANES
1910.180(b)(3)

1(3) *Designated personnel.* Only designated personnel shall be permitted to operate a crane covered by this section.

910.180(1)(5)(ii)

(ii) Operating and maintenance personnel shall be made familiar with the use and care of the fire extinguishers provided.

DERRICKS
1910.181(b)(3)

(3) *Designated personnel.* Only designated personnel shall be permitted to operate a derrick covered by this section.

1910.181(j)(3)(ii)

(ii) Operating and maintenance personnel shall be familiar with the use and care of the fire extinguishers provided.

WOODWORKING MACHINERY REQUIREMENTS
1910.213(s)(5)

(5) Sharpening or tensioning of saw blades or cutters shall be done only by persons of demonstrated skill in this kind of work.

MECHANICAL POWER PRESSES
1910.217(e)(3)

(3) *Training of maintenance personnel.* It shall be the responsibility of the employer to insure the original and continuing com-

petence of personnel caring for, inspecting, and maintaining power presses.

1910.217(f)(2)

(2) *Instruction to operators.* The employer shall train and instruct the operator in the safe method of work before starting work on any operation covered by this section. The employer shall insure by adequate supervision that correct operating procedures are being followed.

FORGING MACHINES

1910.218(a)(2)(iii)

(2) *Inspection and maintenance.* It shall be the responsibility of the employer to maintain all forge shop equipment in a condition which will ensure continued safe operation. This responsibility includes:

(iii) Training personnel for the proper inspection and maintenance of forging machinery and equipment.

WELDING, CUTTING, AND BRAZING

1910.252(b)(1)(iii)

(iii) *Instruction.* Workmen designated to operate arc welding equipment shall have been properly instructed and qualified to operate such equipment as specified in subparagraph (4) of this paragraph.

1910.252(b)(4)(i) and (i)(a)

(i) *General.* Workmen assigned to operate or maintain arc welding equipment shall be acquainted with the requirements of subparagraphs (b), (d), (e), and (f) of this section and, if doing gas shielded arc welding, the if doing gas shielded arc welding, the Recommended Safe Practices for Gas Shielded Arc Welding, AG.1-1966, American Welding Society.

(a) The operator should report any equipment defect or safety hazard to his supervisor and the use of the equipment shall be discontinued until its safety has been assured. Repairs shall be made only by qualified personnel.

1910.252(c)(1)(i) and (iii)

(i) *Installation.* All equipment shall be installed by a qualified electrician in conformance with Subpart S of this part. . . .

(iii) *Personnel.* Workmen designated to operate resistance welding equipment shall have been properly instructed and judged competent to operate such equipment.

(6) *Maintenance.* Periodic inspection shall be made by qualified maintenance personnel, and records of the same maintained. The operator shall be instructed to report any equipment defects to his supervisor and the use of the equipment shall be

discontinued until safety repairs have been completed.

1910.252(d)(2)(iii)(b)

(b) Fire watchers shall have fire extinguishing equipment readily available and be trained in its use. . . .

1910.252(d)(2)(xiii)(c)

(c) Insist that cutters or welders and their supervisors are suitably trained in the safe operation of their equipment and the safe use of the process.

1910.252(f)(13)

(13) *First-aid equipment.* First-aid equipment shall be available at all times. On every shift of welding operations there should be present employees trained to render first aid. . . .

PULP PAPER, AND PAPERBOARD MILLS
1910.261(h)(3)(ii)

(ii) Gas masks capable of absorbing chlorine shall be supplied, conveniently placed, and regularly inspected, and workers who may be exposed to chlorine gas shall be instructed in their use.

SUNDRY MACHINERY AND OPERATIONS
1910.264(d)(1)(v)

(v) *Instruction of employees.* Employees shall be properly instructed as to the hazards of their work and be instructed in safe practices, by bulletins, printed rules, and verbal instructions.

SAWMILLS
1910.265(c)(3)(x)

(x) *Lift trucks.* Lift trucks shall be designed, constructed, maintained, and operated in accordance with the requirements of 1910.178. [Only trained and authorized operators shall be permitted to operate a powered industrial truck. Methods shall be devised to train operators in the safe operation of powered industrial trucks.)

PULPWOOD LOGGING
1910.266(c)(5)(1) and (ii)

(i) Chain saw operators shall be instructed to inspect the saws daily to assure that all handles and guards are in place and tight, that all controls function properly, and that the muffler is operative.

(ii) Chain saw operators shall be instructed to follow manufacturer's instructions as to operation adjustment.

1910.266(c)(6)(i), (vi), (vii), (ix), (x)

(i) Equipment operators shall be instructed as to the manufacturer's recommendations for equipment operation, maintenance, safe practices, and site operating procedures.

(vi) The equipment operator shall be instructed to walk completely around machine and assure that no obstacles or personnel are in the area before startup.

(vii) The equipment operator shall be instructed to start and operate equipment only from the operator's station or from a safe area recommended by the manufacturer.

(ix) The equipment operator shall be instructed to check all controls for proper function and response before starting working cycle.

(x) The equipment operator shall be instructed to pound or secure all movable elements when not in use.

1910.266(c)(6)(xiii) through (xxi)

(xiii) The equipment operator shall be instructed to maintain adequate distance from other equipment and personnel.

(xiv) Where signalmen are used, the equipment operator shall be instructed to operate the equipment only on signal from the designated signalman and only when signal is distinct and clearly understood.

(xv) The equipment operator shall be instructed not to operate movable elements (boom, grapple, load, etc.) close to or over personnel.

(xvi) The equipment operator shall be instructed to signal his intention before operation when personnel are in or near the working area.

(xvii) The equipment operator shall be instructed to dismount and stand clear for all loading and unloading of this mobile vehicle by other mobile equipment. The dismounted operator shall be visible to loader operator.

(xviii) The equipment operator shall be instructed to operate equipment in a manner that will not place undue shock loads on wire rope.

(xix) The equipment operator shall be instructed not to permit riders or observers on the machine unless approved seating and protection is provided.

(xx) The equipment operator shall be instructed to shut down the engine when the equipment is stopped, apply brake locks and ground moving elements before he dismounts.

(xxi) The equipment operator shall be instructed, when any equipment is transported from one job location to another, to transport it on a vehicle of sufficient rated capacity and the equipment shall be properly secured during transit.

1910.266(c)(7)

(7) *Explosives.* Only trained and experienced personnel shall handle or use explosives. Usage shall comply with the requirements of 1910.109.

1910.266(e)(2)(i) and (ii)

(i) The feller shall be instructed to plan retreat path and clear path as necessary before cut is started.

(ii) The feller shall be instructed to appraise situation for dead limbs, the lean of tree to be cut, wind conditions, location of other trees and other hazards and exercise proper precautions before cut is started.

TELECOMMUNICATIONS
1910.268(c)

(c) *Training.* Employers shall provide training in the various precautions and safe practices described in this section and shall insure that employees do not engage in the activities to which this section applies until such employees have received proper training in the various precautions and safe practices required by this section. However, where the employer can demonstrate that an employee is already trained in the precautions and safe practices required by this section prior to his employment, training need not be provided to that employee in accordance with this section. Where training is required, it shall consist of on-the-job training or classroom-type training or a combination of both. The training program shall include a list of the subject courses and the types of personnel required to receive such instruction. A written description of the training program and a record of employees who have received such training shall be maintained for the duration of the employee's employment and shall be made available upon request to the Assistant Secretary for Occupational Safety and Health. Such training shall, where appropriate, include the following subjects:

(1) Recognition and avoidance of dangers relating to encounters with harmful substances, and animal, insect, or plant life.

(2) Procedures to be followed in emergency situations, and

(3) First-aid training, including instruction in artificial respiration.

1910.268(g)(3)(ii)

(ii) The employer shall ensure that pole climbers are inspected by a competent person for the following conditions: Fractured or cracked gaffs or leg irons, loose or dull gaffs, broken straps or buckles. If any of these conditions exist, the defect shall be corrected before the climbers are used.

1910.268(h)(1)

(1) The employer shall ensure that no employee nor any material or equipment may be supported or permitted to be supported on any portion of a ladder unless it is first determined, by inspections and checks conducted by a competent person, that such ladder is adequately strong, in good condition, and properly secured in place, as required in Subpart D of this part and as required in this section.

1910.268(j)(1)(i) and (ii)

(i) The employer shall ensure that visual inspections are made of the equipment by a competent person each day the equipment is to be used to ascertain that it is in good condition.

(ii) The employer shall ensure that tests shall be made at the beginning of each shift by a competent person to insure the vehicle brakes and operating systems are in proper working condition.

(d) Only persons trained in the operation of the derrick shall be permitted to operate the derrick.

(f) The employer shall ensure that the derrick and its associated equipment are inspected by a competent person at intervals set by the manufacturer but in no case less than once per year. Records shall be maintained including the dates of inspections, and necessary repairs made, if corrective action was required.

(1) Employees involved in using high voltages to locate trouble or test cables shall be instructed in the precautions necessary for their own safety, and the safety of other employees.

1910.268(o)(1)(ii), and (o)(3)

(ii) While work is being performed in the manhole, a person with basic first-aid training shall be immediately available to render assistance if there is cause for believing that a safety hazard exists, and if the requirements contained in paragraphs (d)(1) and (o)(1)(i) of this section do not adequately protect the employee(s).

(3) *Joint power and telecommunication manholes.* While work is being performed in a manhole occupied jointly by an electric utility and a telecommunication utility, an employee with basic first-aid training shall be available in the immediate vicinity to render emergency assistance as may be required. The employee whose presence is required in the immediate vicinity for the purposes of rendering emergency assistance is not to be precluded from occasionally entering a manhole to provide assistance other than in an emergency. The requirement of this paragraph does not preclude a qualified employee [any worker who by reason of his training and experience has demonstrated his ability to safely perform his duties], working alone, from entering for brief periods of time, a manhole where energized cables or equipment are in service, for the purpose of inspection, housekeeping, taking readings, or similar work if such work can be performed safely.

1910.268(q)(1)(ii)(a)

(ii) Employees engaged in line-clearing operations shall be instructed that:

(a) A direct contact is made when any part of the body touches or contacts an energized conductor, or other energized electrical fixture or apparatus.

1910.268(q)(1)(ii)(b) through (d)

(b) An indirect contact is made when any part of the body touches any object in contact with an energized electrical conductor, or other energized fixture or apparatus.

(c) An indirect contact can be made through conductive tools, tree branches, trucks, equipment, or other objects, or as a result of communications wires, cables, fences, or guy wires being accidentally energized.

(d) Electric shock will occur when an employee, by either direct or indirect contact with an energized conductor, energized tree limb, tools, equipment, or other object, provides a path for the flow of electricity to a grounded object or to the ground itself. Simultaneous contact with two energized conductors will also cause electric shock which may result in serious or fatal injury.

1910.268(q)(2)(ii)

(ii) Only qualified employees [any worker who by reason of his training and experience has demonstrated his ability to safely perform his duties] or trainees, familiar with the special techniques and hazards involved in line clearance, shall be permitted to perform the work if it is found that an electrical hazard exists.

1910.268(q)(2)(iii) and (iv)

(iii) During all tree working operations aloft where an electrical hazard of more than 750V exists, there shall be a second employee [a tree worker who through related training and on the job experience is familiar with the special techniques and hazards involved in line clearance] or trainee qualified in line clearance trimming [any worker regularly assigned to a line clearance tree trimming crew and undergoing on-the-job training whom in the course of such training, has demonstrated his ability to perform his duties safely at his level of training] within normal voice communication.

(iv) Where tree work is performed by employees qualified in line clearance tree trimming and trainees qualified in line clearance tree trimming, the clearances from energized conductors given in Table R-3 shall apply.

QUALIFICATIONS OF DIVE TEAM
1910.410(a)(1) through (4)

(1) Each dive team member shall have the experience or training necessary to perform assigned tasks in a safe and healthful manner.

(2) Each dive team member shall have experience or training in the following:

(i) The use of tools, equipment, and systems relevant to assigned tasks;

(ii) Techniques of the assigned diving mode; and

(iii) Diving operations and emergency procedures.

(3) All dive team members shall be trained in cardiopulmonary resuscitation and first aid (American Red Cross standard course or equivalent).

(4) Dive team members who are exposed to or control the exposure of others to hyperbaric conditions shall be trained in diving-related physics and physiology.

1910.410(b)(1)

(1) Each dive team member shall be assigned tasks in accordance with the employee's experience or training, except that limited additional tasks may be assigned to an employee undergoing training provided that these tasks are performed under the direct supervision of an experienced dive team member.

(2) The designated person-in-charge shall have experience and training in the conduct of the assigned diving operation.

ASBESTOS

1910.1001(j)(5)

(5) *Employee information and training.*

(i) The employer shall institute a training program for all employees who are exposed to airborne concentrations of asbestos, tremolite, anthophyllite, actinolite, or a combination of these minerals at or above the action level and ensure their participation in the program.

(ii) Training shall be provided prior to or at the time of initial assignment and at least annually thereafter.

(iii) The training program shall be conducted in a manner which the employee is able to understand. The employer shall ensure that each employee is informed of the following:

(A) The health effect associated with asbestos, tremolite, anthophyllite, or actinolite exposure;

(B) The relationship between smoking and exposure to asbestos, tremolite, anthophyllite, and actinolite in producing lung cancer;

(C) The quantity, location, manner of use, release, and storage of asbestos, tremolite, anthophyllite, or actinolite, and the specific nature of operations which could result in exposure to asbestos, tremolite, anthophyllite, or actinolite;

(D) The engineering controls and work practices associated with the employee's job assignments.

(E) The specific procedures implemented to protect employees from exposure to asbestos, tremolite, anthophyllite, or actinolite, such as appropriate work practices, emergency and clean-up procedures, and personal protective equipment to be used;

(F) The purpose, proper use, and limitations of respirators and protective clothing;

(G) The purpose and a description of the medical surveillance program required by paragraph (1) of this section;

(H) A review of this standard, including appendices.

(iv) *Access to information and training materials.*

(A) The employer shall make a copy of this standard and its appendices readily available without cost to all affected employees.

(B) The employer shall provide, upon request, all materials relating to the employee information and training program to the Assistant Secretary and the training program to the Assistant Secretary and the Director.

The (i) and (ii) training requirements listed here pertain to all chemicals immediately shown at left, through N-Nitrosodimethylamine.

N-NITROBIPHENYL
1910.1003(e)(5)(i) and (ii)

ALPHA-NAPHTHYLAMINE
1910.1004(e)(5)(i) and (ii)

METHYL CHLOROMETHYL ETHER
1910.1006(e)(5)(i) and (ii)

3,3′-DICHLOROBENZIDINE (and its salts)
1910.1007(e)(5)(i) and (ii)

BI-CHLOROMETHYL ETHER
1910.1008(e)(5)(i) and (ii)

BETA-NAPHTHYLAMINE
1910.1009(e)(5)(i) and (ii)

BENZIDINE

(i) Each employee prior to being authorized to enter a regulated area, shall receive a training and indoctrination program including, but not necessarily limited to:

(a) The nature of the carcinogenic hazards of 4-Nitrobiphenyl, and others listed at left, including local and systemic toxicity;

(b) The specific nature of the operation involving 4-Nitrobiphenyl which could result in exposure;

(c) The purpose for and application of the medical surveillance program, including, as appropriate, methods of self examination;

(d) The purpose for and application of decontamination practices and purposes;

(e) The purpose for and significance of emergency practices and procedures;

(f) The employee's specific role in emer-

1910.1010(e)(5)(i) and (ii)

4-AMINODIPHENYL

1910.1011(e)(5)(i) and (ii)

gency procedures;

(g) Specific information to aid the employee in recognition and evaluation of conditions and situations which may result in the release of 4-Nitrobiphenyl;

(h) The purpose for and application of specific first aid procedures and practices;

(i) A review of this section at the employee's first training and indoctrination program and annually thereafter.

(ii) Specific emergency procedures shall be prescribed, and posted, and employees shall be familiarized with their terms, and rehearsed in their application.

ETHYLENEIMINE

1910.1012(e)(5)(i) and (ii)

BETA-PROPIOLACTONE

1910.1013(e)(5)(i) and (ii)

2-ACETYLAMINOFLUORENE

1910.1014(e)(5)(i) and (ii)

4-DIMETHYLAMINO-AZOBENZENE

1910.1015(e)(5)(i) and (ii)

N-NITROSODIMETHYLAMINE

1910.1016(e)(5)(i) and (ii)

VINYL CHLORIDE

1910.1017(j)(1)(i) through (ix)

(j) *Training*. Each employee engaged in vinyl chloride or polyvinyl chloride operations shall be provided training in a program relating to the hazards of vinyl chloride and precautions for its safe use.

(1) The program shall include:

(i) The nature of the health hazard from chronic exposure to vinyl chloride including specifically the carcinogenic hazard;

(ii) The specific nature of operations which could result in exposure to vinyl chloride in excess of the permissible limit and necessary protective steps;

(iii) The purpose for, proper use, and limitations of respiratory protective devices;

(iv) The fire hazard and acute toxicity of vinyl chloride, and the necessary protective steps;

(v) The purpose for and a description of the monitoring program;

(vi) The purpose for, and a description of, the medical surveillance program;

(vii) Emergency procedures;

(viii) Specific information to aid the employee in recognition of conditions which may result in the release of vinyl chloride; and

(ix) A review of this standard at the employee's first training and indoctrination program, and annually thereafter.

INORGANIC ARSENIC
1910.1018(o(1) and (2)

(1) *Training program.*

(i) The employer shall institute a training program for all employees who are subject to exposure to inorganic arsenic above the action level without regard to respirator use, or for whom there is a possibility of skin or eye irritation from inorganic arsenic. The employer shall assure that those employees participate in the training program.

(ii) The training program shall be provided by October 1, 1978, for employees covered by this provision, at the time of initial assignment for those subsequently covered by this provision, and shall be repeated at least quarterly for employees who have optional use of respirators and at least annually for other covered employees thereafter, and the employer shall assure that each employee is informed of the following:

(A) The information contained in Appendix A;

(B) The quantity, location, manner of use, storage, sources of exposure, and the specific nature of operations which could result in exposure to inorganic arsenic as well as any necessary protective steps;

(C) The purpose, proper use, and limitation of respirators;

(D) The purpose and a description of the medical surveillance program as required by paragraph (n) of this section;

(E) The engineering controls and work practices associated with the employee's job assignment; and

(F) A review of this standard.

(2) *Access to training materials.*

(i) The employer shall make readily available to all affected employees a copy of this standard and its appendixes.

(ii) The employer shall provide, upon request, all materials relating to the employee information and training program to the Assistant Secretary and the Director.

LEAD
1910.1025(i)(1) and (2)

(i) Each employer who has a workplace in which there is a potential exposure to airborne lead at any level shall inform employees of the content of Appendixes A and B of this regulation.

(ii) The employer shall institute a training program for and assure the participation of all employees who are subject to exposure to lead at or above the action level or for whom the possibility of skin or eye irritation exists.

(iii) The employer shall provide initial training prior to initial job assignment.

(iv) The training program shall be repeated at least annually for each employee.

(v) The employer shall assure that each employee is informed of the following:

(A) The content of this standard and its appendixes;

(B) The specific nature of the operations which could result in exposure to lead above the action level;

(C) The purpose, proper selection, fitting, use, and limitations of respirator;

(D) The purpose and a description of the medical surveillance program, and the medical removal protection program including information concerning the adverse health effects associated with excessive exposure to lead (with particular attention to the adverse reproductive effects on both males and females);

(E) The engineering controls and work practices associated with the employee's job assignment;

(F) The contents of any compliance plan in effect; and

(G) Instructions to employees that collating agents should not routinely be used to remove lead from their bodies and should not be used at all except under the direction of a licensed physician;

(2) *Access to information and training materials.*

(i) The employer shall make readily available to all affected employees a copy of this standard and its appendixes.

(ii) The employer shall provide, upon request, all materials relating to the employee information and training program to the Assistant Secretary and the Director.

(iii) In addition to the information required by paragraph (v), the employer shall include as part of the training program, and shall distribute to employees, any materials pertaining to the Occupational Safety and Health Act, the regulations issued pursuant to that Act, and this lead standard, which are made available to the employer by the Assistant Secretary.

COKE OVEN EMISSIONS
1910.1029(k)(1) and (2)

(1) *Training program.*

(i) The employer shall institute a training program for employees who are employed in the regulated area and shall assure their participation.

(ii) The training program shall be provided as of January 27, 1977 for employees who are employed in the regulated area at that time or at the time of initial assignment to a regulated area.

(iii) The training program shall be provided at least annually for all employees who are employed in the regulated area, except that training regarding the occupational safety and health hazards associated with exposure to coke oven emissions and the purpose, proper use, and limitations of respiratory protective devices shall be provided at least quarterly until January 20, 1978.

(iv) The training program shall include informing each employee of:

(a) The information contained in the substance information sheet for coke oven emissions (Appendix A);

(b) The purpose, proper use, and limitations of respiratory protective devices required in accordance with paragraph (g) of this section;

(c) The purpose for and a description of the medical surveillance program required by paragraph (j) of this section including information on the occupational safety and health hazards associated with exposure to coke oven emissions;

(d) A review of all written procedures and schedules required under paragraph (f) of this section: and

(e) A review of this standard.

(2) *Access to training materials.*

(i) The employer shall make a copy of this standard and its appendixes readily available to all employees who are employed in the regulated area.

(ii) The employer shall provide upon request all materials relating to the employee information and training program to the Secretary and the Director.

COTTON DUST
1910.1043(i)(1) and (2)

(1) *Training program.*

(i) The employer shall provide a training program for all employees in all workplaces where cotton dust is present, and shall assure that each employee in these workplaces is informed of the following:

(a) The specific nature of the operations which could result in exposure to cotton dust at or above the permissible exposure limit;

(b) The measures, including work practices required by paragraph (g) of this section, necessary to protect the employee from exposures in excess of the permissible exposure limit;

(c) The purpose, proper use and limitations of respirators required by paragraph (f) of this section;

(d) The purpose for and a description of the medical surveillance program required by paragraph (h) of this section and other information which will aid exposed employees in understanding the hazards of cotton dust exposures; and

(e) The contents of this standard and its appendixes.

(ii) The training program shall be provided prior to initial assignment and shall be repeated at least annually.

(2) *Access to training materials.*

(i) Each employer shall post a copy of this section with its appendixes in a public location at the workplace, and shall, upon request, make copies available to employees.

(ii) The employer shall provide all materials relating to the employee training and information program to the Assistant Secretary and the Director upon request.

(iii) In addition to the information required by paragraph (i), the employer shall include as part of the training program, and shall distribute to employees, any materials, pertaining to the Occupational Safety and Health Act, the regulations issued pursuant to that Act, and this cotton dust standard, which are made available to the employer by the Assistant Secretary.

1,2-DIBROMO-3-CHLOROPROPANE
1910.1044(n)(1) and (2)

(1) *Training program.*

(i) The employer shall institute a training program for all employees who may be exposed to DBCP and shall assure their participation in such training program.

(ii) The employer shall assure that each employee is informed of the following:

(a) The information contained in Appendix A.

(b) The quantity, location, manner of use, release or storage of DBCP and the specific nature of operations which could result in exposure to DBCP as well as any necessary protective steps;

(c) The purpose, proper use, and limitations of respirators;

(d) The purpose and description of the medical surveillance program required by paragraph (m) of this section; and

(e) A review of this standard, including appendixes.

(2) *Access to training materials.*

(i) The employer shall make a copy of this standard and its appendixes readily available to all affected employees.

(ii) The employer shall provide, upon request, all materials relating to the employee information and training program to the Assistant Secretary and the Director.

ACRYLONITRILE (VINYL CYANIDE)
1910.1045(o)(1) and (2)

(1) *Training program.*

(i) By January 2, 1979, the employer shall institute a training program for and assure the participation of all employees exposed to AN above the action level, all employees whose exposures are maintained below the action level by engineering and work practice controls, and all employees subject to potential skin or eye contact with liquid AN.

(ii) Training shall be provided at the time of initial assignment, or upon institution of the training program, and at least annually thereafter, and the employer shall assure that each employee is informed of the following:

(A) The information contained in Appendixes A and B. *Editor's Note*: See Federal Register, Vol. 43, No. 192, Oct. 3, 1978, pp. 45813–45815;

(B) The quantity, location, manner of use, release, or storage of AN, and the specific nature of operations which could result in exposure to AN, as well as any necessary protective steps:

(C) The purpose, proper use, and limitations of respirators and protective clothing;

(D) The purpose and a description of the medical surveillance program required by paragraph (n) of this section;

(E) The emergency procedures developed, as required by paragraph (i) of this section;

(F) Engineering and work practice controls, their function, and the employee's relationship to these controls; and

(G) A review of this standard.

(2) *Access to training materials.*

(i) The employer shall make a copy of this standard and its appendixes readily available to all affected employees.

(ii) The employer shall provide, upon request, all materials relating to the employee information and training program to the Assistant Secretary and the Director.

ETHYLENE OXIDE
1910.1047(j)(3)(iii)

(3) *Information only training.*

(i) The employer shall provide employees who are potentially exposed to EtO at or above the action level with information and training on EtO at the time of initial assignment and at least annually thereafter.

(ii) Employees shall be informed of the following:

(A) The requirements of this section with an explanation of its contents, including Appendices A and B;

(B) Any operations in their work area where EtO is present;

(C) The location and availability of the written EtO final rule; and

(D) The medical surveillance program required by paragraph (i) of this section with an explanation of the information in Appendix C.

(iii) Employee training shall include at least:

(A) Methods and observations that may be used to detect the presence or release of EtO in the work area (such as monitoring conducted by the employer, continuous monitoring devices, etc.);

(B) The physical and health hazards of EtO;

(C) The measures employees can take to protect themselves from hazards associated with EtO exposure, including specific procedures the employer has implemented to protect employees from exposure to EtO, such as work practices, emergency procedures, and personal protective equipment to be used; and

(D) The details of the hazard communication program developed by the employer, including an explanation of the labeling system and how employees can obtain and use the appropriate hazard information.

HAZARD COMMUNICATION
1910.1200(h)(2)

(2) *Training.* Employee training shall include at least:

(i) Methods and observations that may be used to detect the presence or release of a hazardous chemical in the work area (such as monitoring conducted by the employer, continuous monitoring devices, visual appearance or odor of hazardous chemicals when being released, etc.);

(ii) The physical and health hazards of the chemicals in the work area;

(iii) The measures employees can take to protect themselves from these hazards, including specific procedures the employer has implemented to protect employees from exposure to hazardous chemicals, such as appropriate work practices, emergency procedures, and personal protective equipment to the used; and,

(iv) The details of the hazard communication program developed by the employer, including an explanation of the labeling system and the material safety data sheet, and how employees can obtain and use the appropriate hazard information.

## CONSTRUCTION INDUSTRY TRAINING REQUIREMENTS

The following training requirements have been excerpted from Title 29, Code of Federal Regulations Part 1926. Note that in addition to these training requirements, Part 1910, relating to general industry, also contains applicable trianing standards.

| Subject and Standard Number | Training Requirement |
|---|---|
| GENERAL SAFETY AND HEALTH PROVISIONS<br>1926.20(b)(2) and (4) | (2) Such programs [as may be necessary to comply with this part] shall provide for frequent and regular inspections of the job sites, materials, and equipment to be made by competent persons [capable of identifying existing and predictable hazards in the suroundings or working conditions which are unsanitary, hazardous, or dangerous to employees, and who have authorization to take prompt corrective measures to eliminate them] designated by the employers.<br>(4) The employer shall permit only those employees qualified [one who by possession of a recognized degree, certificate, or professional standing, or who by extensive knowledge, training, and experience, has successfully demonstrated his ability to solve or resolve problems relating to the subject matter, the work, or the project] by training or experience to operate equipment and machinery. |
| SAFETY TRAINING AND EDUCATION<br>1926.21(a)<br>1926.21(b)(1) through (6) | (a) *General requirements.* The Secretary shall, pursuant to section 107(f) of the Act, establish and supervise programs for the education and training of employers and employees in the recognition, avoidance and prevention of unsafe conditions in employment covered by the Act.<br>(1) The employer should avail himself of the safety and health training programs the Secretary provides.<br>(2) The employer shall instruct each employee in the recognition and avoidance of unsafe conditions and the regulations applicable to his work environment to control or eliminate any hazards or other exposure to illness or injury.<br>(3) Employees required to handle or use poisons, caustics, and other harmful substances shall be instructed regarding their safe handling and use, and be made aware |

of the potential hazards, personal hygiene, and personal protective measures required.

(4) In job site areas where harmful plants or animals are present, employees who may be exposed shall be instructed regarding the potential hazards and how to avoid injury and the first aid procedures to be used in the event of injury.

(5) Employees required to handle or use flammable liquids, gases, or toxic materials shall be instructed in the safe handling and use of these materials and made aware of the specific requirements contained in Subparts D, F, and other applicable subparts of this part.

(6) (i) All employees required to enter into confined or enclosed spaces shall be instructed as to the nature of the hazards involved, the necessary precautions to be taken, and in the use of protective and emergency equipment required. The employer shall comply with any specific regulations that apply to work in dangerous or potentially dangerous areas.

(6) (ii) For purposes of subdivision (i) of this subparagraph, "confined or enclosed space" means any space having a limited means of egress, which is subject to the accumulation of toxic or flammable contaminants or has an oxygen deficient atmosphere. Confined or enclosed spaces include, but are not limited to, storage tanks, process vessels, bins, boilers, ventilation or exhaust ducts, sewers, underground utility vaults, tunnels, pipelines, and open top spaces more than 4 feet in depth such as pits, tubs, vaults, and vessels.

MEDICAL SERVICES AND FIRST AID
1926.50(c)

(c) In the absence of an infirmary, clinic, hospital, or physician that is reasonably accessible in terms of time and distance to the worksite which is available for the treatment of injured employees, a person who has a valid certificate in first-aid training from the U.S. Bureau of Mines, the American Red Cross, or equivalent training that can be verified by documentary evidence, shall be available at the worksite to render first aid.

IONIZING RADIATION
1926.53(b)

(b) Any activity which involves the use of radioactive materials or X-rays, whether or not under license from the Atomic Energy Commission [Nuclear Regulatory Commission] shall be performed by competent persons specially trained in the proper and safe operation of such equipment. In the case of materials used under Commission license, only persons actually licensed, or competent persons under the direction and supervision of the licensee, shall perform such work.

NONIONIZING RADIATION
1926.54(a) and (b)

(a) Only qualified and trained employees shall be assigned to install, adjust and operate laser equipment.

(b) Proof of qualification of the laser equipment operator shall be available and in possession of the operator at all times.

GASES, VAPORS, FUMES, DUSTS, AND MISTS
1926.55(b)

(b) To achieve compliance with paragraph (a) of this section, administrative or engineering controls must first be implemented whenever feasible. When such controls are not feasible to achieve full compliance, protective equipment or other protective measures shall be used to keep the exposure of employees to air contaminants within the limits prescribed in this section. Any equipment and technical measures used for this purpose must first be approved for each particular use by a competent industrial hygienist or other technically qualified person. Whenever respirators are used, their use shall comply with 51926.103.

ASBESTOS, TREMOLITE, ANTHOPHYLLITE, AND ACTINOLITE
1926.58(k)(3)

(3) *Employee information and training.* (i) The employer shall institute a training program for all employees exposed to airborne concentrations of asbestos, tremolite, anthophyllite, actinolite, or a combination of these minerals in excess of the action level and shall ensure their participation in the program.

(ii) Training shall be provided prior to or at the time of initial assignment, unless the employee has received equivalent training within the previous 12 months, and at least annually thereafter.

(iii) The training program shall be conducted in a manner that the employee is able to understand. The employer shall ensure that each such employee is informed of the following:

(A) Methods of recognizing asbestos, tremolite, anthophyllite, and actinolite;

(B) The health effects associated with asbestos, tremolite, anthophyllite, or actinolite exposure;

(C) The relationship between smoking and asbestos, tremolite, anthophyllite, and actinolite in producing lung cancer;

(D) The nature of operations that could result in exposure to asbestos, tremolite, anthophyllite, or actinolite, the importance of necessary protective controls to minimize exposure including, as applicable, engineering controls, work practices, respirators, housekeeping procedures, hygiene facilities, protective clothing, decontamination procedures, emergency procedures, and waste disposal procedures, and any necessary instruction in the use of these controls and procedures;

(E) The purpose, proper use, fitting instructions, and limitations of respirators as required by 29 CFR 1910.134;

(F) The appropriate work practices for performing the asbestos, tremolite, anthophyllite, or actinolite job; and

(G) Medical surveillance program requirements.

(H) A review of this standard, including appendices.

(4) *Access to training materials.* (i) The employer shall make readily available to all affected employees without cost all written materials relating to the employee training program, including a copy of this regulation.

(ii) The employer shall provide to the Assistant Secretary and the Director, upon request, all information and training materials relating to the employee information and training program.

HEARING PROTECTION
1926.101(b)

(b) Ear protective devices inserted in the ear shall be fitted or determined individually by competent persons.

RESPIRATORY PROTECTION
1926.103(c)(1)

(1) Employees required to use respiratory protective equipment approved for use in atmospheres immediately dangerous to life shall be thoroughly trained in its use. Employees required to use other types of respiratory protective equipment

shall be instructed in the use and limitations of such equipment.

FIRE PROTECTION
1926.150(a)(5)

(5) As warranted by the project, the employer shall provide a trained and equipped fire fighting organization (Fire Brigade) to assure adequate protection to life. "Fire brigade" means an organized group of employees that are knowledgeable, trained, and skilled in the safe evacuation of employees during emergency situations and in assisting in fire fighting operations.

1926.150(c)(1)(viii)

(viii) Portable fire extinguishers shall be inspected periodically and maintained in accordance with Maintenance and Use of Portable Fire Extinguishers, NFPA No. 10A1970.

*From ANSI Standard 10A-1970.* The owner or occupant of a property in which fire extinguishers are located has an obligation for the care and use of these extinguishers at all times. By doing so, he is contributing to the protection of life and property. The nameplate(s) and instruction manual should be read and thoroughly understood by all persons who may be expected to use extinguishers.

"1120. To discharge this obligation he should give proper attention to the inspection, maintenance, and recharging of this fire protective equipment. lie should also train his personnel in the correct use of fire extinguishers on the different types of fires which may occur on his property.

"3020. Persons responsible for performing maintenance operations come from three major groups:

"Trained industrial safety or maintenance personnel.

"Extinguisher service agencies.

"Individual owners (e.g., self emloyed . . .)."

SIGNALING
1926.201(a)(2)

(2) Signaling directions by flagmen shall conform to American National Standards Institute D6.1-1971, Manual on Uniform Traffic Control Devices for Streets and Highways.

POWER-OPERATED HAND TOOLS
1926.302(e)(1) and (12)

(1) Only employees who have been trained in the operation of the particular tool in use shall be allowed to operate a power-actuated tool.

(12) Power-actuated tools used by employees shall meet all other applicable requirements of American National Standards Institute, A10.3.1970, Safety Requirements for Explosive-Actuated Fastening Tools.

WOODWORKING TOOLS
1926.304(f)

(f) *Other requirements.* All woodworking tools and machinery shall meet other applicable requirements of American National Standards Institute, 01.1-1961, Safety Code for Woodworking Machinery.

*From ANSI Standard 01.1-1961, Selection and Training of Operators.* Before a worker is permitted to operate any woodworking machine, he shall receive instructions in the hazards of the machine and the safe method of its operation. Refer to A9.7 of the Appendix. "A9.7 Selection and Training of Operators. Operation of Machines, Tools, and Equipment. General."

"(1) Learn the machine's applications and limitations, as well as the specific potential hazards peculiar to this machine. Follow available operating instructions and safety rules carefully.

"(2) Keep working area clean and be sure adequate lighting is available.

"(3) Do not wear loose clothing, gloves, bracelets, necklaces, or ornaments. Wear face, eye, ear, respiratory, and body protection devices, as indicated for the operation or environment.

"(4) Do not use cutting tools larger or heavier than the machine is designed to accommodate. Never operate a cutting tool at greater speed than recommended.

"(5) Keep hands well away from saw blades and other cutting tools. Use a push stock or push block to hold or guide the work when working close to cutting tool.

"(6) Whenever possible, use properly locked clamps, jig, or vise to hold the work.

"(7) Combs (feather boards) shall be provided for use when an applicable guard cannot be used.

"(8) Never stand directly in line with a horizontally rotating cutting tool. This is particularly true when first starting a new tool, or a new tool is initially installed on the arbor.

"(9) Be sure the power is disconnected from the machine before tools are serviced.

"(10) Never leave the machine with the power on.

"(11) Be positive that hold-owns and anti-kickback devices are positioned properly, and that the workpiece is being fed through the cutting tool in the right direction.

"(12) Do not use a dull, gummy, bent, or cracked cutting tool.

"(13) Be sure that keys and adjusting wrenches have been removed before turning power on.

"(14) Use only accessories designed for the machine.

"(15) Adjust the machine for minimum exposure of cutting tool necessary to perform the operation."

GAS WELDING AND CUTTING
1926.350(d)(1) through (6)

(d) *Use of fuel gas.* The employer shall thoroughly instruct employees in the safe use of fuel gas as follows:

(1) Before a regulator to a cylinder valve is connected, the valve shall be opened slightly and closed immediately. (This action is generally termed "cracking" and is intended to clear the valve of dust or dirt that might otherwise enter the regulator.) The person cracking the valve shall stand to one side of the outlet, not in front of it. The valve of a fuel gas cylinder shall not be cracked where the gas would reach welding work, sparks, flame, or other possible sources of ignition.

(2) The cylinder valve shall always be opened slowly to prevent damage to the regulator. For quick closing, valves on fuel gas cylinders shall not be opened more than 1½ turns. When a special wrench is required, it shall be left in position on the stem of the valve while the cylinder is in use so that the fuel gas flow can be shut off quickly in case of an emergency. In the case of manifolded or coupled cylinders, at least one such wrench shall always be available for immediate use. Nothing shall be placed on top of a fuel gas cylinder, when in use, which may damage the safety device or interfere with the quick closing of the valve.

(3) Fuel gas shall not be used from cylinders through torches or other devices which are equipped with shutoff valves without reducing the pressure through a suitable regulator attached to the cylinder valve or manifold.

(4) Before a regulator is removed from a cylinder valve, the cylinder valve shall always be closed and the gas released from the regulator.

(5) If, when the valve on a fuel gas cylinder is opened, there is found to be a leak around the valve stem, the valve shall be closed and the gland nut tightened. If this action does not stop the leak, the use of the cylinder shall be discontinued, and it shall he properly tagged and removed from the work area. In the event that fuel gas should leak from the cylinder valve, rather than from the valve stem, and the gas cannot be shot off, the cylinder shall be properly tagged and removed from the work area. If a regulator attached to a cylinder valve will effectively stop a leak through the valve seat, the cylinder need not be removed from the work area.

(6) If a leak should develop at a fuse plug or other safety device, the cylinder shall be removed from the work area.

1926.350(j)

(j) *Additional rules.* For additional details not covered in this subpart, applicable technical portions of American National Standards Institute, 249.1-1967, Safety in Welding and Cutting, shall apply.

*From ANSI Standard 249.1-1967, Fire Watch Duties.* Fire watchers shall be trained in the use of fire extinguishing equipment. They shall be familiar with facilities for sounding an alarm in the event of a fire. They shall watch for fires in all exposed areas, try to extinguish them only when obviously within the capacity of the equipment available, or otherwise sound the alarm. A fire watch shall be maintained for at least a half hour after completion of welding or cutting operations to detect and extinguish possible smoldering fires."

ARC WELDING AND CUTTING
1926.351(d)(1) through (5)

(d) *Operating instructions.* Employers shall instruct employees in the safe means of arc welding and cutting as follows:

(1) When electrode holders are to be left

unattended, the electrodes shall be removed and the holders shall be so placed or protected that they cannot make electrical contact with employees or conducting objects.

(2) Hot electrode holders shall not be dipped in water; to do so may expose the arc welder or cutter to electric shock.

(3) When the arc welder or cutter has occasion to leave his work or to stop work for any appreciable length of time, or when the arc welding or cutting machine is to be moved, the power supply switch to the equipment shall be opened.

(1) Any faulty or defective equipment shall be reported to the supervisor.

(5) Other requirements, as outlined in Article 630, National Electrical Code, NFPA 70-1971; ANSI C1-1971 (Rev. of 1968), Electric Welders, shall be used when applicable.

FIRE PREVENTION
1926.352(e)

(e) When the welding, cutting, or heating operation is such that normal fire prevention precautions are not sufficient, additional personnel shall be assigned to guard against fire while the actual welding, cutting, or heating operation is being performed, and for a sufficient period of time after completion of the work to ensure that no possibility of fire exists. Such personnel shall be instructed as to the specific anticipated fire hazards and how the fire fighting equipment provided is to be used.

WELDING, CUTTING AND HEATING IN WAY OF PRESERVATIVE
1926.354(a)

(a) Before welding, cutting, or heating is commenced on any coatings surface covered by a preservative coating whose flammability is not known, a test shall be made by a competent person to determine its flammability. Preservative coatings shall be considered to be highly flammable when scrapings burn with extreme rapidity.

GROUND-FAULT PROTECTION
1926.404(b)(iii)(B)

(iii)(B) The employer shall designate one or more competent persons as defined in 51926.32(f) to implement the program.

SCAFOLDING
1926.451(a)(3)

(3) No scaffold shall be erected, moved, dismantled, or altered except under the supervision of competent persons.

1926.451(b)(16)

(16) All wood pole scaffolds 60 feet or less in height shall be constructed and erected in accordance with Tables IA to 10. If they are over 60 feet in height, they shall be

designed by a qualified engineer competent in this field, and it shall be constructed and erected in accordance with such design.

1926.451(c)(4) and (5) (4) Tube and coupler scaffolds shall be limited in heights and working levels to those permitted in Tables L-10, 11, and 12. Drawings and specifications of all tube and coupler scaffolds above the limitations in Tables S-10, 11, and 12 shall be designed by a qualified engineer competent in this field.

(5) All tube and coupler scaffolds shall be constructed and erected to support four times the maximum intended loads, as set forth in Tables S-10, 11, and 12, or as set forth in the specifications by a licensed professional engineer competent in this field.

1926.451(d)(9) (9) Drawings and specifications for all frame scaffolds over 125 feet in height above the base plates shall be designed by a registered professional engineer.

1926.451(g)(3) (3) Unless outrigger scaffolds are designed by a registered professional engineer competent in this field, they shall be constructed and erected in accordance with Table L-13. Outrigger scaffolds, designed by a registered professional engineer, shall be constructed and erected in accordance with such design.

1926.451(h)(6) and (14) (h) *Masons' adjustable multiple-point suspension scaffolds.*

(6) Where the overhang exceeds 6 feet 6 inches, outrigger beams shall be composed of stronger beams or multiple beams and be installed under the supervision of a competent person.

(14) Each scaffold shall be installed or relocated under the supervision of a competent person.

1926.451(k)(10) (k) *Single-point adjustable suspension scaffolds.*

(10) For additional details not covered in this paragraph, applicable technical portions of American National Standards Institute, A120.1-1970, Power-operated Devices for Exterior Building Maintenance Powered Platforms, shall be used.

From ANSI Standard A120.1-1970, "Qualified Operators. Powered platforms shall be operated only by qualified persons who have been instructed in the operation and in the inspection, with respect to safe operating conditions of the particular powered platform to be operated."

## GUARDING OF LOW-PITCHED ROOF PERIMETERS DURING THE PERFORMANCE OF BUILT-UP IN ROOFING WORK

1926.500(g)(6)

(6) *Training.* (i) The employer shall provide a training program for all employees engaged in built-up roofing work so that they are able to recognize and deal with the hazards of falling associated with working near a roof perimeter. The employees shall also be trained in the safety procedures to be followed in order to prevent such falls.

(ii) The employer shall assure that employees engaged in built-up roofing work have been trained and instructed in the following areas:

(a) nature of fall hazards in the work area near a roof edge;

(b) The function, use, and operation of the MSS system, warning line, and safety monitoring systems to be used;

(c) The correct procedures for erecting, maintaining, and disassembling the systems to be used;

(d) The role of each employee in the safety monitoring system when this system is used;

(e) The limitations on the use of mechanical equipment; and

(P) The correct procedures for the handling and storage of equipment and materials.

(iii) Training shall be provided for each newly hired employee, and for all other employees as necessary, to assure that employees maintain proficiency in the areas listed in paragraph (g)(6) (ii) of this section.

## CRANES AND DERRICKS

1926.550(a)(1), (5), and (6)

(1) The employer shall comply with the manufacturer's specifications and limitations applicable to the operation of any and all cranes and derricks. Where manufacturer's specifications are not available, the limitations assigned to the equipment shall be based on the determinations of a qualified engineer competent in this field and such determinations will be appropriately documented and recorded. Attachments used with cranes shall not exceed the capacity, rating, or scope recommended by the manufacturer.

(5) The employer shall designate a competent person who shall inspect all machinery and equipment prior to each use, and during use, to make sure it is in safe operating condition. Any deficiencies shall be repaired, or defective parts replaced, before continued use.

(6) A thorough, annual inspection of the hoisting machinery shall be made by a competent person, or by a government or private agency recognized by the U.S. Department of Labor. The employer shall maintain a record of the dates and results of inspections for each hoisting machine and piece of equipment.

MATERIAL HOISTS, PERSONNEL HOISTS, AND ELEVATORS
1926.552(a)(1)

(1) The employer shall comply with the manufacturer's specifications and limitations applicable to the operation of all hoists and elevators. Where manufacturer's specifications are not available, the limitations assigned to the equipment shall be based on the determinations of a professional engineer competent in the field.

1926.552(b)(7)

(7) All material hoist towers shall be designed by a licensed professional engineer.

1926.552(c)(15) and (17)(i)

(c) *Personnel hoists.*

(15) Following assembly and erection of hoists, and before being put in service, an inspection and test of all functions and safety devices shall be made under the supervision of a competent person. A similar inspection and test is required following major alteration of an existing installation. All hoists shall be inspected and tested at not more than 3-month intervals. Records shall be maintained and kept on file for the duration of the job.

(17)(i) Personnel hoists used in bridge tower construction shall be approved by a registered professional engineer and erected under the supervision of a qualified engineer competent in this field.

MATERIAL HANDLING EQUIPMENT
1926.602(c)(1)(vi)

(c) *Lifting and hauling equipment* (other than equipment covered under Subpart N of this part). (1)(vi) All industrial trucks in use shall meet the applicable requirements of design, construction, stability, inspection, testing, maintenance, and operation, as defined in American National Standards Institute B16.1-1969, Safety Standards for Powered Industrial Trucks.

From ANSI Standard B56.1-1969, "Operator Training. Only trained and authorized operators shall be permitted to operate a powered industrial truck. Methods shall be devised to train operators in the safe operation of powered industrial trucks. Badges or other visual indication of the operators' authorization should be dispiayed at all times during the work period."

SITE CLEARING
1926.604(a)(1)

(1) Employees engaged in site clearing shall be protected from hazards of irritant and toxic plants and suitably instructed in the first aid treatment available.

GENERAL PROTECTION REQUIREMENTS (EXCAVATIONS, TRENCHING, AND SHORING)
1926.650(i)

(i) Daily inspections of excavations shall be made by a competent person. If evidence of possible cave-ins or slides is apparent, all work in the excavation shall cease until the necessary precautions have been taken to safeguard the employees.

SPECIFIC EXCAVATION REQUIREMENTS
1926.651(d), (f), (k), (o), and (x)

(d) Excavations shall be inspected by a competent person or after every rainstorm or other hazard-increasing occurrence, and the protection against slides and cave-ins shall be increased if necessary.

(f) Supporting systems, i.e., piling, cribbing, shoring, etc., shall be designed by a qualified person and meet accepted engineering requirements or practices which are compatible with standards required by a registered architect, a registered professional engineer, or other duly licensed or recognized authority. . . .

(k) Support systems shall be planned and designed by a qualified person when excavation is in excess of 20 feet in depth, adjacent to structures or improvements, or subject to vibration or ground water.

(o) If the stability of adjoining buildings or walls is endangered by excavations, shoring, bracing, or underpinning shall be provided as necessary to insure their safety. Such shoring, bracing, or underpinning shall be inspected daily or more often, as conditions warrant, by a competent person and the protection effectively maintained.

(x) Where ramps are used for employees or equipment, they shall be designed and constructed by qualified persons in accordance with accepted engineering requirements.

FORMS AND SHORING
1926.701(a)(4)

(4) Imposition of any construction loads on the partially completed structure shall not be permitted unless such loading has been considered in the design and approved by the engineer-architect.

1926.701(d)(3)

(3) Whenever single post shores are used in more than one tier, the layout shall be designed and inspected by a structural engineer.

BOLTING, RIVETING, FITTING-UP, AND PLUMBING-UP
1926.752(d)(4)

(4) Plumbing-up guys shall be removed only under the supervision of a competent person.

TUNNELS AND SHAFTS
1926.800(c)(2)(iii) and (iv)

(iii) When there has been a failure of ventilation and ventilation has been restored in a reasonable time, all places where flammable gas may have accumulated shall be examined by a competent person and determined to be free of flammable gas before power is restored and work resumed.

(iv) When the main fan or fans have been shut down with all employees out of the drift, tunnel, or shaft, no employee, other than those qualified to examine the drift, tunnel, or shaft, or other authorized employee, shall go underground until the fans have been started, the work areas examined for gas and other hazards, and declared safe.

1926.800(e)(1)(xii)

(xii) At tunnel operations, employing 25 or more employees at one time underground, at least two rescue crews (10 employees divided between shifts) shall be trained annually in rescue procedures, in the use, care, and limitations of oxygen breathing apparatus, and the use and maintenance of fire fighting equipment. Not less than one crew (5 employees) shall be trained at smaller operations.

1926.800(k)(1)

(1) Equipment that is to be used during a shift shall be inspected by a competent person each shift. Equipment defects affecting safety shall be corrected before the equipment is used.

COMPRESSED AIR
1926.803(a)(1) and (2)

(1) There shall be present, at all times, at least one competent person designated by and representing the employer, who shall be familiar with this subpart in all respects, and responsible for full compliance with these and other applicable subparts.

(2) Every employee shall be instructed

in the rules and regulations which concern his safety or the safety of others.

1926.803(b)(1) and (10)(xii)

(1) There shall be retained one or more licensed physicians familiar with and experienced in the physical requirements and the medical aspects of compressed air work and the treatment of decompression illness. He shall be available at all times while work is in progress in order to provide medical supervision of employees employed in compressed air work. He shall himself be physically qualified and be willing to enter a pressurized environment.

(10) The medical lock shall: (xii) Be in constant charge of an attendant under the direct control of the retained physician. The attendant shall be trained in the use of the lock and suitably instructed regarding steps to be taken in the treatment of employee exhibiting symptoms compatible with a diagnosis of decompression illness.

1926.803(e)(1)

(1) Every employee going under air pressure for the first time shall be instructed on how to avoid excessive discomfort.

1926.803(h)(1)

(1) At all times there shall be a thoroughly experienced, competent, and reliable person on duty at the air control valves as a gauge tender who shall regulate the pressure in the working areas. During tunneling operations, one gauge tender may regulate the pressure in not more than two headings: Provided that the gauge and controls are all in one location. In caisson work, there shall be a gauge tender for each caisson.

PREPARATORY OPERATIONS
1926.850(a)

(a) Prior to permitting employees to start demolition operations, an engineering survey shall be made, by a competent person, of the structure to determine the condition of the framing, floors, and walls, and possibility of unplanned collapse of any portion of the structure. Any adjacent structure where employees may be exposed shall also be similarly checked. The employer shall have in writing evidence that such a survey has been performed.

CHUTES
1926.852(c)

(c) A substantial gate shall be installed in each chute at or near the discharge end. A competent employee shall be assigned to control the operation of the gate, and the backing and loading of trucks.

MECHANICAL DEMOLITION
1926.859(g)

(9) During demolition, continuing inspections by a competent person shall be made as the work progresses to detect hazards resulting from weakened or deteriorated floors, or walls, or loosened material. No employee shall be permitted to work where such hazards exist until they are corrected by shoring, bracing, or other effective means.

GENERAL PROVISIONS (BLASTING AND USE OF EXPLOSIVES)
1926.900(a)

(a) The employer shall permit only authorized and qualified persons to handle and use explosives.

1926.900(k)(3)(i)

(i) The prominent display of adequate signs, warning against the use of mobile radio transmitters, on all roads within 1,000 feet of blasting operations. Whenever adherence to the 1,000-foot distance would create an operational handicap, a competent person shall be consulted to evaluate the particular situation, and alternative provisions may be made which are adequately designed to prevent any premature firing of electric blasting caps. A description of any such alternatives shall be reduced to writing and shall be certified as meeting the purposes of this subdivision by the competent person consulted. The description shall be maintained at the construction site during the duration of the work, and shall be available for inspection by representatives of the Secretary of Labor.

1926.900(q)

(q) All loading and firing shall be directed and supervised by competent persons thoroughly experienced in this field.

BLASTER QUALIFICATIONS
1926.901(c), (d), and (e)

(c) A blaster shall be qualified, by reason of training, knowledge, or experience, in the field of transporting, storing, handling, and use of explosives, and have a working knowledge of State and local laws and regulations which pertain to explosives.

(d) Blasters shall be required to furnish satisfactory evidence of competency in handling explosives and performing in a safe manner the type of blasting that will be required.

(e) The blaster shall be knowledgeable and competent in the use of each type of blasting method used.

SURFACE TRANSPORTATION OF EXPLOSIVES
1926.902(b) and (i)

(b) Motor vehicles or conveyances transporting explosives shall only be driven by, and be in charge of, a licensed driver who is physically fit. He shall be familiar with the local, State, and Federal regulations governing the transportation of explosives.

(i) Each vehicle used for transportation of explosives shall be equipped with a fully charged fire extinguisher, in good condition. An Underwriters Laboratory-approved extinguisher of not less than 10 ABC rating will meet the minimum requirement. The driver shall be trained in the use of the extinguisher on his vehicle.

FIRING THE BLAST
1926.909(a)

(a) A code of blasting signals equivalent to Table U-1, shall be posted on one or more conspicuous places at the operation, and all employees shall be required to familiarize themselves with the code and conform to it. Danger signs shall be placed at suitable locations.

**Table U-L**

WARNING SIGNAL—A 1-minute series of long blasts 5 minutes prior to blast signal.
BLAST SIGNAL—A series of short blasts 1 minute prior to the shot.
ALL CLEAR SIGNAL—A prolonged blast following the inspection of blast area.

GENERAL REQUIREMENTS (POWER LINES AND TRANSMISSION AND DISTRIBUTION)
1926.950(d)(1)(ii), (vi), and (vii)

(1) When de-energizing equipment operated in excess of 600 volts, and the means of disconnecting from electric energy is not visibly open or visibly locked out, the provisions of subdivisions (i) through (vii) of this subparagraph shall be complied with:

(ii) Notification and assurance from the designated employee [a qualified person delegated to perform specific duties under the conditions existing] shall be obtained that:

(a) All switches and disconnectors through which electric energy may be supplied to the particular section of line or equipment to be worked have been de-energized;

(b) All switches and disconnectors are plainly tagged indicating that men are at work;

(c) And that where design of such switches and disconnectors permits, they have been rendered inoperable.

(vi) When more than one independent crew requires the same line or equipment to be de-energized, a prominent tag for each such independent crew shall be placed on the line or equipment by the designated employee in charge.

(vii) Upon completion of work on de-energized lines or equipment, each designated employee in charge shall determine that all employees in his crew are clear, that protective grounds installed by his crew have been removed, and he shall report to the designated authority that all tags protecting his crew may be removed.

1926.950(d)(2)(ii) (2) When a crew working on a line or equipment can clearly see that the means of disconnecting from electric energy are visibly open or visibly locked-out, the provisions of subdivisions (i) and (ii) of this subparagraph shall apply:

(ii) Upon completion of work on de-energized lines or equipment, each designated employee in charge shall determine that all employees in his crew are clear, that protective grounds installed by his crew have been removed, and he shall report to the designated authority that all tags protecting his crew may be removed.

1926.950(e)(1) and (2) (1) The employer shall provide training or require that his employees are knowledgeable and proficient in:

(i) Procedures involving emergency situations, and

(ii) First-aid fundamentals including resuscitation.

(2) In lieu of subparagraph (1) of this paragraph the employer may comply with the provisions of 51926.5Hc regarding first-aid requirements.

OVERHEAD CRANES

1926.955(b)(3)(1) (3)(1) A designated employee shall be used in directing mobile equipment adjacent to footing excavations.

1926.955(b)(8) and (d)(i) (8) A designated employee shall be utilized to determine that required clearance is maintained in moving equipment under or near energized lines.

(1) Prior to stringing parallel to an existing energized transmission line a compe-

tent determination shall be made to ascertain whether dangerous induced voltage buildups will occur, particularly during switching and ground fault conditions. When there is a possibility that such dangerous induced voltage may exist the employer shall comply with the provisions of subparagraphs (2) through (9) of this paragraph in addition to the provisions of paragraph (c) of this § 1926.955, unless the line is worked as energized.

(1) Employees shall be instructed and trained in the live-line bare-hand technique and the safety requirements pertinent thereto before being permitted to use the technique on energized circuits.

(4) All work shall be personally supervised by a person trained and qualified to perform live-line bare-hand work.

UNDERGROUND LINES
1926.956(b)(1)

(1) While work is being performed in manholes, an employee shall be available in the immediate vicinity to render emergency assistance as may be required. This shall not preclude the employee in the immediate vicinity from occasionally entering a manhole to provide assistance, other than emergency. This requirement does not preclude a qualified employee [a person who by reason of experience or training is familiar with the operation to be performed and the hazards involved], working alone, from entering for brief periods of time, a manhole where energized cables or equipment are in service, for the purpose of inspection, housekeeping, taking readings, or similar work if such work can be performed safely.

CONSTRUCTION IN ENERGIZED SUBSTATIONS
1926.957(a)(1)

(1) When construction work is performed in an energized substation, authorization shall be obtained from the designated, authorized person [a qualified person delegated to perform specific duties under the conditions existing] before work is started.

1926.957(d)(1)

(1) Work on or adjacent to energized control panels shall be performed by designated employees.

1926.957(e)(1)

(1) Use of vehicles, gin poles, cranes, and other equipment in restricted or hazardous areas shall at all times be controlled by designated employees.

## SUMMARY

OSHA training requirements are obviously numerous and complex. They also change frequently. Only a constant review of recent OSHA changes can keep your health and safety training current.

Additionally, many industries or processes may be subject to additional requirements from other state and/or federal and/or local regulatory agencies.

# CHAPTER 4

# TRAINING TECHNOLOGY

## INTRODUCTION

To meet the demands and challenges of the ever changing workplace and the ever increasing need for improved productivity and efficiency, the modern training program must take advantage of opportunities for automation.

Available and emerging technology can greatly facilitate the administration of a training program by automating registration, scheduling, and documentation.

Additionally, there is automated assistance to aid in the development of training programs, the production of course materials and audiovisual aids, course presentation, and training evaluation.

The use of available technology is highly recommended for all safety training efforts and is really mandatory for large, complex programs.

## USING TECHNOLOGY FOR SAFETY AND HEALTH TRAINING ADMINISTRATION

### Registration

There are a number of specialized software packages to aid in registering attendees for classes or courses. Additionally, it is relatively simple to set up a program for entering attendees, dates, courses, classrooms, and instructors into any good data base software program and then generating course rosters and other training reports by simply using the sorting and report generation capabilities of the software package.

In larger organizations it may be desirable to keep a "course catalog" or listing of available training on a central computer (mainframe or server on a local area network). The calendar of scheduled courses can be kept on the same system.

In large organizations with complex training requirements and frequently changing schedules, this is the only practical way of ensuring everyone has instant access to the latest data.

Many programs allow direct electronic registration, eliminating or greatly reducing the need for a "paper" system.

### Scheduling

Many scheduling tasks can also be accomplished using standard data base software. Again there are specialized programs for training administration and management and for scheduling.

Good automated programs can eliminate "double booking" of classrooms, classroom support equipment, or instructors and can ensure the efficient use of available resources.

There are many ways in which technology can aid in the safety and health training effort. One example is given in Figure 4.1. This type of document can be posted in the training facility or provided to students in advance to aid in locating the proper classroom for third class. It can also serve as a training calendar to let managers, supervisors, employees, and trainers know what, when, and where training is being offered. Finally this same form can be used by the classroom coordinator as a scheduling tool.

### Training Records

Documentation of training has always been important for audit and tracking purposes and as more organizations adopt an "if it isn't documented, it didn't happen" philosophy, the need for accurate, timely, auditable individual and organizational training records increases.

Paper systems can document training by the use of sign-in sheets and/or class rosters.

Many current "automated" systems simply take this data and manually enter it into a data base for storage, manipulation, retrieval, and report generation purposes.

More sophisticated systems are integrated so that initial registration for the course provides the necessary data for attendance documentation and the only additional data entry may be the deletion of "no shows" from the data base. Proof or verification of attendance can be provided by collecting individual tests or other evaluation tools or course critiques and cross-checking number of attendees and names with the electronic records. Other verification alternatives include bar codes on employee badges and bar code scanners, thumb or fingerprint readers, or other electronic recognition devices.

TRAINING CALENDAR: April 10 TO April 14

| Room No. | MONDAY | TUESDAY | WEDNESDAY | THURSDAY | FRIDAY |
|---|---|---|---|---|---|
| 1 | OSHA WORKSHOP → | | | | → |
| 2 | GENERAL EMPLOYEE ORIENTATIONS → | | | | → |
| 3 | LOCKOUT/ TAGOUT | CONFINED SPACES → | | → | LOCKOUT/ TAGOUT |
| 4 | HAZCOM → | → | | HAZCOM → | → |
| 5 | SUPERVISORS SAFETY TRAINING | | | LOCKOUT/ TAGOUT | |
| 6 | ACCIDENT INVESTIGATION → | | | | → |
| 7 | HAZCOM → | → | SAFETY OBSERVER TRAINING | | |
| 8 | ROOT CAUSE ANALYSIS → | | | | → |
| 9 | | | SUPERVISORS SAFETY TRAINING | | |
| 10 | SYSTEM SAFETY WORKSHOP → | | | | → |

**FIGURE 4.1** Training calendar/classroom schedule.

## Tracking

The tracking of individual training progress or the monitoring of organizational training status can be done very easily and simply by reports generated from standard or specialized training data base programs.

Any good relational data base can sort and report training records many ways for many purposes. The type of report shown in Figure 4.2 would be beneficial in identifying individuals to serve on an accident board of investigation.

Individual reports can document in-house and outside education and training, complete with dates, locations, instructors, and test scores. These reports can also indicate what refresher and/or additional education or training is required or desired, when and where that training may be available, and registration information (or, in a fully automated system, the computer may preregister attendees for needed training and generate a memo or letter advising the attendee of the time and place to attend).

Summary reports can be generated for any level of the organization to provide managers or supervisors with the training status of their personnel. These reports can indicate the percentage of personnel who have attended specific courses or who have or have not completed required training. They can also indicate when additional or refresher training is due and/or scheduled.

| | ACCIDENT INVESTIGATION WORKSHOP | | | |
|---|---|---|---|---|
| NAME | DATE | LOCATION | INSTRUCTOR | |
| Adams, A. A. | 02/'15/95 | Las Vegas, NV | Seidel, S.S. | |
| Brown, B. B. | 04/22/95 | Chicago, IL | Watts, W. W. | |
| Clark, C. C. | 06/10/95 | Jackson, MS | Stephenson,, S. S. | |
| Davis, D. D. | 02/15/95 | Las Vegas, NV | Seidel, S.S. | |
| Evans, E. E. | 02/15/95 | Las Vegas, NV | Seidel, S.S. | |
| Fluggenhaumer, F. F. | 04/22/95 | Chicago, IL | Watts, W. W. | |
| Goosestein, G. G. | 02/15/95 | Las Vegas, NV | Seidel, S.S. | |
| Hicks, H. H. | 06/10/95 | Jackson, MS | Stephenson, S. S. | |
| Ivy, I. I. | 06/10/95 | Jackson, MS | Stephenson, S. S. | |
| Jacobsbury, J. J. | 02/15/95 | Las Vegas, NV | Seidel, S.S. | |
| Kilroy, K.K. | 04/22/95 | Chicago, IL | Watts, W. W. | |

**FIGURE 4.2** Course training summary.

## USING TECHNOLOGY FOR COURSE DEVELOPMENT

### Job and Task Analysis

Standard wordprocessing, data base, and spreadsheet software may offer some assistance in systematically analyzing jobs to determine what tasks must be performed to do the job and then analyzing the tasks to determine the knowledge and skills required to perform each task.

Specialized software is also available and much more helpful. Unfortunately, it is relatively expensive.

### Audiovisual Aids

The array of equipment and software available to aid in producing audiovisual aids is mind-boggling and limited only by budget.

Standard graphics and presentation software make the production of overhead transparencies a snap. Black and white transparencies can be printed directly on standard laser printers or produced by copying paper originals onto transparency film on plain paper copiers. Color transparencies obviously require a color printer or plotter.

Instant 35-mm slides can also be produced using computer generated graphics and various types of cameras or recorders. There are online services available that can transfer computer generated graphics electronically (by modem) to vendors who will prepare color slides and mail them to the originator. Computer-generated graphics and animation can also be transferred to video.

The production of high-quality audiovisual aids used to be time consuming and costly. Now, computer-generated graphics and huge libraries of clipart (on a single disk or CD-ROM) make the production of audiovisual aids quick and inexpensive (Figure 4.3).

Both standard video camcorders and still video cameras can be used to produce audiovisual aids to be shown on standard or large screen television sets or monitors or projected using video projectors. Still pictures taken with standard cameras can also be transferred to video.

## Examinations

Standard word processing software can be extremely helpful in preparing, updating, rearranging, storing, and printing written examinations.

More sophisticated specialized software can also aid in phrasing and formatting questions, automatically generating tests from lists of questions, automatically generating various versions of the same test, and performing item analyses of responses to evaluate specific questions and responses. Some of this software also provides test result storage and tracking and trending capabilities.

Scanning equipment can grade tests, summarize and analyze results, and store and print reports in minutes (or seconds).

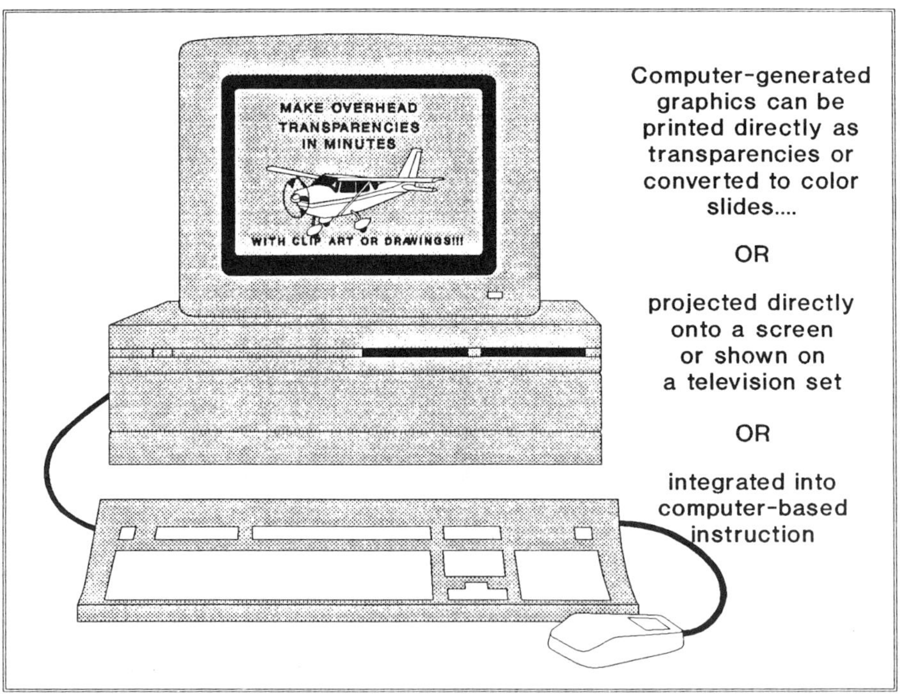


**FIGURE 4.3** Computer-generated overhead.

## USING TECHNOLOGY FOR COURSE PRESENTATION

### Video

The most common use of video in the classroom is to show "off-the-shelf" video productions. This is a very cost effective way of presenting information in a concise and interesting way. A good video production is approximately 4 times as efficient as the average standup lecture. That is, information that normally takes an hour to present in a lecture format can be presented in about 15 minutes on video tape.

Video presentations can be used effectively for audiences ranging from the individual trainee to very large audiences using multiple monitors and/or projection television and/or closed circuit or even broadcast television.

In addition to simply playing off-the-shelf programs, in-house productions can provid tailored job and site-specific information.

Cloud circuit or satellite feed telecasts can aid in providing interactive, live instruction to audiences in multiple locations.

Videocassettes can be a very important part of self-study courses providing individualized instruction.

In-house video productions are especially effective for presenting environment, safety, and health instruction. Actual video footage of job sites and processes showing details of operations and highlighting potential hazards are invaluable tools for teaching hazard recognition and presenting job orientations. Video footage of accidents can frequently be turned into timely, highly effective tools for showing the workforce what went wrong and how the accident could have been avoided.

### Computer-Based Training

Computer-based training (CBT) can be a very cost-effective method of presenting material to individuals. The interactive nature of CBT provides self-pacing, which means that all trainees can operate at their appropriate learning rate. Fast learners or persons already familiar with the subject can move very rapidly while slower learners or those totally unfamiliar with the material can move at an equally challenging but slower rate.

Computer-based training is a "natural" for teaching computer skills. Computer-based tutorials are available for most software packages. Many are included as an integral part of the software. Even though it is desirable to have "live" help available in the form of a subject matter expert, the best way to learn to use many software packages is by sitting down at the computer and working through computer-based tutorials, supplemented by the written users manuals and the "help" features built into the software.

Computer-based programs can also be linked to video cassette players, laser disks, and/or CD-ROM players to provide "moving pictures" to the programs (see Interactive Video, below). As the sophistication of computer-generated animation improves, the importance of alternate means of providing "motion" to the computer screen may be diminished.

Computer-based training is an excellent choice for situations where a "live" instructor may not be practical, such as for new employee orientations in small organizations where only a few new employees are processed daily or weekly. Other good target audiences for computer-based training are managers, professionals, and other "highly compensated" employees where it is very desirable to have training continuously available so that it can be conducted at the convenience of the trainee.

Another great advantage of computer-based instruction is that it can automatically record and track individual trainee progress. A computer-based training program can administer and grade examinations and can store the results. It can also provide instant feedback and remedial training. Many programs can also evaluate learning rates and provide "report cards."

### Interactive Video

Interactive video is a type of computer-based training that combines video programs (on video cassettes, laser discs, CD-ROMs, or computer discs) with a computer to provide self-paced, interactive instruction. Interaction may be accomplished using a computer keyboard but many modern interactive programs use touch screen monitors to receive trainee input and record trainee responses.

Like other computer-based programs, interactive video programs may be capable of storing and evaluating training data to include dates and times of training sessions, test scores, learning rates, etc.

A shortcoming of interactive video is that hardware and software are relatively expensive, tending to limit its application to relatively small audiences and making it difficult to provide and maintain sufficient hardware to train large target audiences in a short time frame.

### Other

Other technology that can aid in the presentation of training includes

- Devices to project computer screen output onto large screens via an overhead projector.
- Electronic reader boards that can continuously present short safety and health awareness messages and brief training items in the workplace, break areas, or cafeterias.
- Large "white boards" that can electronically produce hard copies of items written on the board. These are great for developing ideas or concepts on the board or recording trainee responses. The thoughts or responses captured on the board can instantly be preserved and distributed to attendees

without having to manually copy the material off the board. This allows all concerned to concentrate on the subject without having to divert attention to note taking.

- Control devices for audiovisuals. These range from very sophisticated controllers that can "present" multiscreen, multimedia shows to simple remote controls for video players, TV monitors, and slide projectors.
- Simulators, video games, and virtual reality (it is sometimes hard to tell where "video games"—especially the arcade versions—stop and "serious" simulator/trainers begin) all offer opportunities to learn now to safely drive automobiles and motor cycles, fly planes, operate equipment, and learn/practice other tasks in a very safe environment (albeit the focus of most arcade games is *not* "safe operation" and I not sure that all video arcades are, in fact, safe environments).

The technology to practice emergency procedures and to learn to operate construction equipment, complex manufacturing equipment, and perform other potentially hazardous jobs in a variety of simulated conditions and environments already exists. The potential applications in the areas of safety training and operator certification are almost unlimited.

## USING TECHNOLOGY FOR PROGRAM EVALUATION

### Tracking and Trending

The relative effectiveness of individual courses and/or instructors can be tracked and trended by entering examination results, course critiques, and/or follow-up evaluations into data base or spreadsheet programs.

Similarly individual or organizational trends can be detected and tracked by examining training data bases to evaluate training attendance and performance.

Additionally, evaluation instruments (examinations, quizzes, performance evaluations, etc.) can be evaluated by performing automated item analyses and by examining trends in performance on these instruments over time.

### Performance Indicators and Measures

Automated training data bases can be sorted to provide performance indicators and measures such as

- The number of student hours of instruction provided per month by organization or by individual instructor
- Classroom utilization

- Instructor utilization
- The cost of instruction by course or by student hour
- Average attendance by course or by attending organization
- Total attendance by course or by month
- Relationship between safety and health training and accident rates (to include the statistical significance of the relationship).
- Relationship between safety and health training costs and the costs of accidents (to include the statistical significance of the relationship).

Relevant performance indicators should be periodically charted and communicated to the organization in order to provide feedback or performance.

Most graphics, spreadsheet, and data base software packages include the capability to produce graphs and charts. Figure 4.4 gives an example of a safety training performance indicator graph.

As we discussed here and in Chapter 5 on recordkeeping and in Chapter 13 on evaluation of training, computers and available software make the tracking and trending of safety training performance indicators and related records and data fast, efficient, and (done properly) effective. Figure 4.4 shows the monthly "production" of four safety instructors. Care should be exercised in

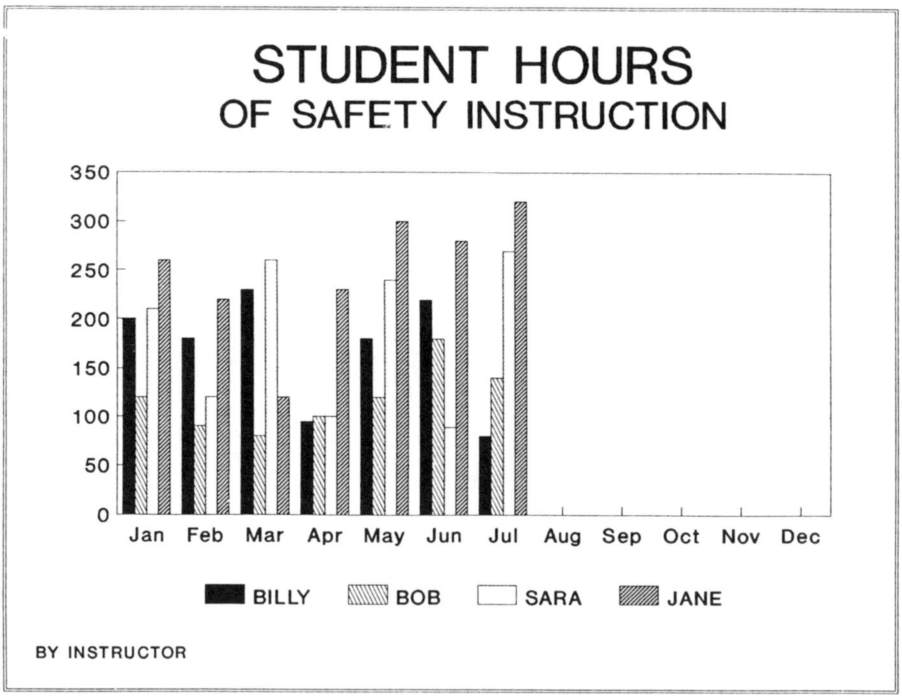


**FIGURE 4.4** Student hours of instruction/by instructor.

in interpreting this type of information. For example, the information on this chart does not indicate that Jane is the "best" or the "most valuable" safety trainer. Jane may do little development work and may teach one or two "basic" courses to large audiences whereas Bob may have the primary responsibility for developing new courses and may teach several advanced or specialized courses or workshops to relatively small audiences.

## WHERE TO GET HELP

With the rapid advances in technology, especially computer software, it is probably not realistic to attempt to recommend software packages. (The word processing software used to prepare this manuscript has been upgraded three times in the last 18 months. The hardware and software used to prepare the first edition of *Safety Training Methods* are so "archaic," we had to retype or scan hardcopy before we could edit and revise.)

You should need no help in identifying basic word processing, data base, spreadsheet, or graphics/presentation packages.

Computer-based multimedia courses, however, which can be highly interactive and include high quality sound, text, graphics, animation and full-motion video—all on the same computer disc or CD-ROM—are still relatively new.

Authoring software to create this type of program includes the latest versions of older, well-established DOS-based programs such as Allen Communications Quest and Computer Teaching Corporation's TenCORE Language Authoring System.

Some of the earlier multimedia authoring packages for use with Microsoft Windows include Authorware Professional for Windows from Authorware, IconAuthor from Aimtech, and Multimedia ToolBook from Asymetrix. Again, software and even software-producing companies change rapidly as the state-of-the-art advances and companies in this explosive market are created, merged, renamed, and realigned.

To help stay abreast of these rapid and often significant changes, the following partial list of educational technology and computer publications may be of interest (unfortunately, this list is also subject to change):

Educational Technology
Educational Technology Publication Inc.
720 Palisade Ave.
Englewood Cliff, NJ 07632
(201) 871-4007

Electronic Learning
Scholastic Inc.
730 Broadway
New York, NY 10003-9538
(314) 636-5271

PC Week
Ziff-Davis Publishing Co.
1 Park Ave.
New York, NY 10016
(212) 503-3500

T H E Journal (Technological Horizons in Education)
Information Synergy Inc.
P.O. Box 17239
Irvine, CA 92713
(714) 730-4011

Technology & Learning
Peter Li Inc.
24511 E. River Rd.
Dayton, OH 45439
(513) 294-5785

Much of the above information was obtained from the Interactive '93 Conference CD, which was provided to attendees at the Ziff Institute's 1993 Interactive '93 Conference in Anaheim, California. This CD was produced jointly by Ziff Institute and Computer Library and contains articles from approximately 90 publications. Key word searches of over 80,000 articles are completed in less than 3 minutes.

## SUMMARY

Computer, video equipment, and other technology can provide invaluable assistance in administering, developing, presenting, and evaluating training. The wise use of available technology provides improved efficiency and effectiveness. In all but the very smallest organizations, the use of this technology is no longer an option. Leading edge operations and organizations must also be on the leading edge of the application of training technology.

# CHAPTER 5

# SAFETY TRAINING RECORDS

## INTRODUCTION

There are two primary sets of safety and health training records that must be maintained: first, the safety training organization must maintain internal records of trainer qualifications and course and program information, and second, and probably even more important, individual training records must be maintained to document OSHA and other required training. Ideally, these records will document all training, mandatory and nonmandatory, including training provided by outside or external vendors and conferences. Additionally, the company level training records should include organizational summaries of training by organizational unit and by course or program.

Accurate, complete, and auditable documentation of training's essential to ensuring that personnel are, in fact, receiving required training and in planning and scheduling courses.

Accident investigators and OSHA inspectors will look at training documentation to provide an indicator of how well the organization is meeting its safety training requirements.

The type of training record shown in Figure 5.1 can be produced periodically or on demand. It can be distributed in "hard copy" or electronically. An ideal system will have data available to the appropriate people at all times and will always be current and accurate. This is achievable but not always easy. For example, records can be updated as students are registered so that entries appear automatically on the date of course completion with no further input. This means, of course, that canceling a course automatically prevents the course from appearing on training records and that students who fail to complete the course are also deleted.

**INDIVIDUAL TRAINING RECORD**

Smith, Sam S.
Occupational Safety
Mail Stop 333

| COURSE | DATE | REFRESHER DUE |
|---|---|---|
| General Employees Orientation | 03/12/90 | NA |
| Supervisors Safety Course | 04/02/90 | |
| | 03/20/92 | |
| | 04/22/94 | 05/22/96 |
| Defensive Driving | 11/11/90 | NA |
| OSHA Workshop | 03/20/91 | NA |
| Accident Investigation | 06/18/91 | NA |
| Advanced Word Processing | 08/10/91 | NA |
| Introduction to Industrial Hygiene | 12/10/91 | NA |
| Root Cause Analysis | 05/22/92 | NA |
| Hoisting and Rigging Workshop | 07/14/92 | NA |
| Total Quality Management | 10/10/92 | |
| | 10/20/93 | 11/20/96 |
| Introduction to System Safety | 11/14/93 | NA |
| Desktop Publishing (Basic) | 11/22/93 | NA |
| Facility System Safety | 02/18/94 | NA |
| Desktop Publishing (Advanced) | 03/11/94 | NA |
| Environmental Compliance Workshop | 06/12/94 | 07/12/96 |

**FIGURE 5.1** Individual training record.

## TRAINING ORGANIZATION RECORDS

### Qualification Documentation

#### *Assigned Personnel*
The safety training organization should maintain individual training records on all assigned trainers. These records should include all training received as a member of the organization and all other relevant education and training. It is recommended that current resumes be maintained on the training staff, especially if the training organization may be required to prepare and submit internal or external training proposals.

#### *Subject Matter Experts*
The safety and health training organization should also maintain complete qualification records on subject matter experts that assist in developing, presenting, and/or evaluating training.

#### *Adjunct Faculty*
It is also desirable to obtain and maintain resumés of instructors from outside the company.

#### *Schedules*
It is recommended that course/class schedules be maintained in order to provide key information for audit, appraisal, and planning purposes.

#### *Course Critiques*
Course critiques should also be retained and recorded again in order to provide backup documentation for audits or inspections and to provide evaluation and trending data.

#### *Performance Indicators*
Performance indicator charts and data, generated on a weekly, monthly, or quarterly basis, should be maintained in order to conduct long-term studies and provide a basis for projecting performance and establishing goals for continuous improvement.

Figure 5.2 is an example of a performance indicator chart that can be generated from student critiques to provide input to instructors and to training managers.

## COMPANY TRAINING RECORDS

### Individual Training Records

Individual training records should be maintained for all members of the organization. At a minimum, these records should accurately document the

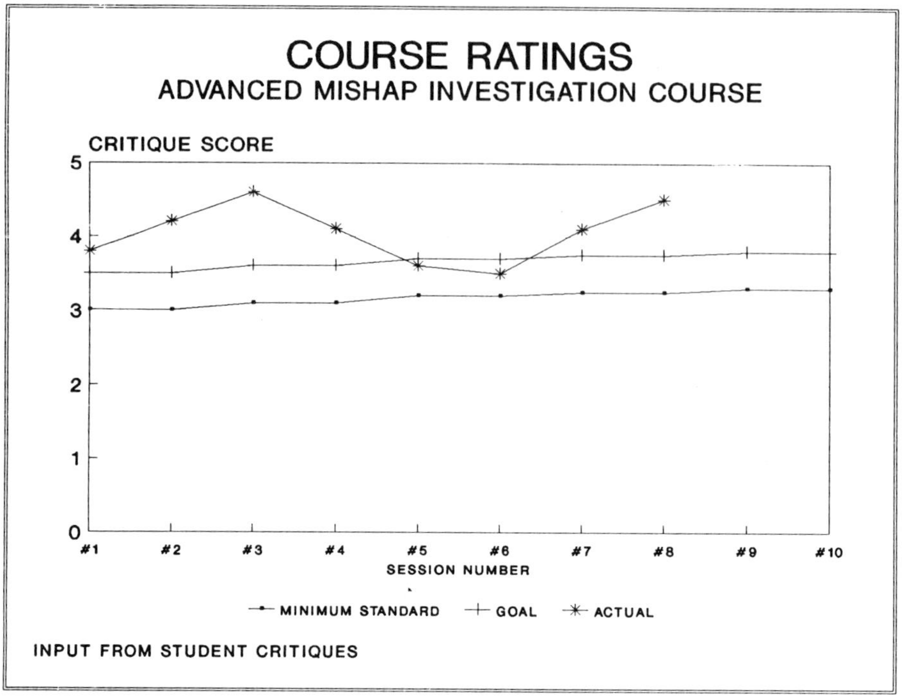


**FIGURE 5.2** Course ratings.

time, place, course title, course length, exam grades (if applicable), and instructor for mandatory training (training required by OSHA or other local, state, or federal code, standard, or regulation and other training designated as mandatory by management). Ideally, individual training records will also include all education and training activities.

These records can be maintained in hard copy and/or as electronic records and should be readily accessible to the individual employee, his manager, and authorized auditors, and inspectors.

## Organizational Summaries

### *Mandatory Training*

Periodic organizational summaries of mandatory training should be provided to responsible managers. There records should the current training status of organizational personnel and should indicate when additional and/or refresher training is due. Ideally, these records should also show when and where required training is available.

These summary reports should also be available on request. Electronic reports should be continuously available to authorized individuals.

### *Developmental Training*

Organizational training summaries should also be provided for skill enhancement and personal development training.

Again, these summaries should be available on individual courses or programs and should be provided periodically and on request.

### *Outside Training*

Records for training provided by outside organizations or vendors should include a summary of the training, the cost of the training, an evaluation of the training value and effectiveness, and recommendations as to appropriate audiences.

Despite a reluctance to mention specific software since it changes so rapidly, there is one recordkeeping package worthy of note. *OSHA Training Manager* by Compliance Solutions International (San Diego, California, 92127, Phone: 619-673-8420, Fax: 619-451-8975) is an excellent tool for small and medium size organizations. The amazing thing about this well designed, user friendly product is that it sells (as of mid-1994) for about $90. In addition, it is designed to link to add-on modules including a Regulatory Training Requirements Reference Guide and a testing module.

## SUMMARY

The accurate and timely maintenance of training records is an essential safety and health training function. These records must be readily accessible to the organization and to audit and oversight groups. These records must be able to document the safety and health training effort and provide accurate and timely feedback as to the status of the safety and health training effort.

## CHAPTER 6

# GENERAL SAFETY TRAINING TOPICS

### INTRODUCTION

In addition to the other topics covered in this section, there are some important general topics that should be included in virtually all safety and health training programs. Three of these topics are

- Employee rights and responsibilities under the Occupational Safety and Health Act
- General housekeeping
- Use of personal protective equipment

### OSHA RIGHTS AND RESPONSIBILITIES

Generally the Occupational Safety and Health Act states that every employee has a right to a safe and healthful workplace and that every employee has a responsibility to comply with applicable safety and health codes, standards, and regulations.

Employees must know their rights and responsibilities under the provisions of the Occupational Safety and Health Act. This information should be integrated into new employee safety and health orientations and safety meetings and should be posted in the workplace (Figure 6.1).

It should be noted that OSHA requirements may not apply to employees of certain federal agencies that have the statutory authority to prescribe or enforce safety and health standards. OSHA requirements also do no apply to farmers with 10 or fewer employees, work activity in a recreational, hunting, or

EMPLOYEE RIGHTS
AND RESPONSIBILITIES

- Employee Rights under OSHA
  - Participate in Standards Process
    (Request standards, testify, comment, object, appeal variances, etc.
  - Have Access to Safety Information
    (Access to NIOSH studies and lists, OSHA rules, safety statistics, exposure records, etc.)
  - Participate in Enforcement Activities
    (Confer with inspectors, attend briefings, request special inspections, file complaints, testify, contest abatement actions, etc.)
- Employee Responsibilities
  - Comply with Safety Rules

**FIGURE 6.1** OSHA rights and responsibilities.

fishing area, or employers with 10 or fewer employees in industries with 3 digit Standard Industrial Classification injury/illness rates of less than 7 per 100 employees. Church workers and domestic household workers are also generally not covered.

Safety and Health training for supervisors and managers should include employer rights and responsibilities.

Safety and Health training for all employees should include employee rights and responsibilities.

### Employer Responsibilities

- Each employer has the general duty to furnish each of his employees employment and places of employment that are free from recognized hazards that are causing or likely to cause death or serious physical harm (this is commonly known as the "general duty clause").
- Each employer has the specific duty of complying with health and safety standards promulgated under the Act.

### Employer Rights

An employer has the right to

- Seek advice and off-site consultation as needed by writing, calling, or visiting the nearest OSHA office.
- Request and receive proper identification of the OSHA compliance officer prior to inspection.

- Be advised by the compliance safety and health officer (CSHO) of the reason for an inspection.
- Have an opening and closing conference with the CSHO.
- File a Notice of Contest with the OSHA area director within 15 working days of receipt of a notice of citation and proposed penalty.
- Apply to OSHA for a temporary variance from a standard if unable to comply because of the unavailability of materials, equipment, or personnel need to make necessary changes within the required time.
- Take an active role in developing safety and health standards through participation in OSHA Standards Advisory Committees, through nationally recognized standards-setting organizations, and through evidence and views presented in writing or at hearings.
- Avail himself, if a small business employer, of long-term loans through the Small Business Administration to help bring the establishment into compliance, either before of after an OSHA inspection.
- Be assured of the confidentiality of any trade secrets observed by an OSHA compliance officer.

## Employee Responsibilities

Each employee has the duty to comply with the safety and health standards and all rules, regulations, and orders that are applicable to his or her actions or conduct on the job.

## Employee Rights

### *With Respect to Standards*

- Employees may request OSHA to begin proceeding for adoption of a new standard or to amend or revoke an existing one.
- Employees may submit written data of comments on proposed standards and may appear as an interested party at any hearing held by OSHA.
- Employees may file written objections to a proposed federal standard and/or appeal the final decision of OSHA.
- Employees must be informed when an employer applies for a variance of a promulgated standard.
- Employees must be afforded the opportunity to participate in a variance hearing as an interested party and have the right to appeal OSHA's final decision.

### *With Respect to Access to Information*

- Employees have the right to information from the employer regarding employee protections and obligations under the Act and to review appropriate OSHA standards, rules, regulations, and requirements that the employer should have available in the workplace.

- Affected employees have a right to information from the employer regarding the toxic effects, conditions of exposure, and precautions for safe use of all hazardous materials in the establishment by means of labeling or other forms of warning where such information is prescribed by a standard.
- If employees are exposed to harmful materials in excess of levels set by the standards, the affected employees must be so informed by the employer and the employer must also inform the employees thus exposed what corrective action is being taken.
- If a compliance safety and health officer determines that an alleged imminent danger exists, he or she must inform the affected employees of the danger and that he or she is recommending that relief be sought by court action if the imminence of such danger is not eliminated.
- Upon request, employees must be given access to records of their history of exposure to toxic materials or harmful physical agent that are required to be monitored or measured and recorded.
- If a standard requires monitoring or measuring hazardous materials or harmful physical agents, employees must be given the opportunity to observe such monitoring or measuring.
- Employees have the right of access to (a) the list of toxic materials published by NIOSH, (b) criteria developed by NIOSH describing the effects of toxic materials or harmful physical agents, and (c) industrywide studies conducted by NIOSH regarding the effects of chronic, low-level exposure to hazardous materials.
- On written request to NIOSH, employees have the right to obtain the determination of whether of not a substance found or used in the establishment is harmful.
- Upon request, the employees should be allowed to review the Log and Summary of Occupational Injuries (OSHA No. 200) at a reasonable time and in a reasonable manner.

### *With Respect to Enforcement*

- Employees have the right to confer in private with the compliance officer and to respond to questions from the compliance officer, in connection with an inspection of an establishment.
- An authorized employee representative must be given an opportunity to accompany the compliance officer during an inspection for the purpose of aiding such inspection. (This is commonly known as the "walk-around" provision.) Also, an authorized employee has the right to participate in the opening and closing conferences during the inspection.
- An employee has the right to make a written request to OSHA for a special inspection if the employee believes a violation of a standard threatens physical harm, and the employee has the right to request OSHA to keep his or her identity confidential.

- If an employee believes any violation of the Act exists, he or she has the right to notify OSHA or a compliance officer in writing of the alleged violation, either before or during an inspection of the establishment.
- If a request is made for a special inspection and it is denied by OSHA, the employee must be notified in writing by OSHA, together with the reasons, that the complaint was not valid. The employee has the right to object to such a decision and may request a hearing by OSHA.
- If a written complaint concerning an alleged violation is submitted to OSHA and the compliance officer responding to the complaint fails to cite the employer for the alleged violation, OSHA must furnish the employee or his or her authorized representative a written statement setting forth the reasons for its final disposition.
- If OSHA cites an employer for a violation, employees have the right to review a copy of the citation which must be posted by the employer at or near the place where the violation occurred.
- Employees have the right to appear as an interested party or to be called as a witness in a contested enforcement matter before the Occupational Safety and Health Review Commission.
- If OSHA arbitrarily or capriciously fails to seek relief to counteract an imminent danger and an employee is injured as a result, that employee has the right to bring action against OSHA for relief as may be appropriate.
- Employees have the right to file a complaint to OSHA within 30 days if they believe they have been discriminated against because they asserted his rights under the Act.
- An employee has the right to contest the abatement period fixed in the citation issued to his or her employer by notifying the OSHA Area Director that issued the citation within 15 working days of the issuance of the citation.

## HOUSEKEEPING

What types of accidents can be prevented through conscientious housekeeping? Here are a few of the more common ones:

- Tripping over loose objects on floors, stairs, and platforms
- Slipping on wet, greasy, or dirty floors
- Bumping against projecting or misplaced material
- Puncturing or scratching hands or other parts of the body on protruding nails, hooks, or rods
- Injuries from falling objects

Unfortunately there are no shortcuts to good housekeeping. Employees should be reminded daily, and one of the most effective reminders is to set a good example. If supervisors keep their own offices or work areas looking neat and uncluttered, their campaigns will be more effective.

### Regular Procedures

Any department will have fewer hazards and will be more efficient if regular procedures for housekeeping are established.

- Set aside 5 minutes at the end of the day for each person to clean his or her own work area. Heavier cleaning chores can be handled in longer periods at the end of each week. Also, intensive cleanups and inspections should be scheduled either on a quarterly, midyear, or year-end basis, depending on the conditions in each department.
- One technique is to post weekly cleanup ratings for each employee, calling attention to shortcomings and recognizing those who are conscientious. Also, in monthly safety meetings, make housekeeping a part of the permanent agenda.

### Places to Concentrate

Floor scrubbing, painting, vacuuming, waxing, and other heavy cleaning chores are usually done by janitorial or maintenance crews. It is management's responsibility, however, to be sure that they are performed satisfactorily. This still leaves a big housekeeping job for each supervisor and his or her workers. Here are some areas to concentrate on:

- **Work Areas.** Mark them clearly with painted lines. Plan production carefully so excessive amounts of work materials or finished items do not clutter work areas.
- **Machines and Equipment.** Arrange machines, benches, tables, and other equipment to combine maximum production with enough room for safe, uncrowded work stations. Provide racks or containers for tools, jigs, and fixtures. Keep benches and tables in good repair and cleanly painted.
- **Aisles.** Be sure they are marked clearly and are wide enough to allow easy traffic flow. Keep them free of material, finished parts, and scrap.
- **Floors.** Check to see that they are vacuumed and scrubbed regularly. Protect floors by providing sawdust or other absorbent material on which to place oily parts. Have oil and grease spots removed from the floor immediately.
- **Walls and Ceilings.** Be sure they are scrubbed and painted when neces-

sary. Provide racks for hanging supplies and clothes. A bulletin board will keep calendars, posters, and cartoons off the walls.

- Other checkpoints include the following:

  Clean, uncluttered stairs

  Adequate storage and care of rags and waste

  Adequate storage and proper handling of oils, solvents, and other flammables

  Effective cleanup methods or procedures

  Adequate storage of raw stock

  Adequate project storage

  Clean light fixtures and windows

  Effective tool storage methods

  Adequate dust and fume removal equipment.

## FACTORS AFFECTING HOUSEKEEPING

The total working environment is the key to housekeeping and housekeeping-related problems. For example, root causes of housekeeping problems are usually inadequate storage space and/or poor workplace layout or organization.

The total working environment must be under constant scrutiny because of changing conditions, new employees, equipment additions and modifications, and so on. The following checklist is presented as a guide to identify potential problems.

### Building

- Correct ceiling height
- Correct floor type; in acceptable condition
- Adequate illumination
- Adequate plumbing and heating pipes and equipment
- Windows with acceptable opening, closing, and holding devices; protection from breakage
- Acceptable size doors with correct swing and operational quality
- Adequate railings and nonslip treads on stairways and balconies
- Adequate ventilation
- Adequate storage facilities
- Adequate electrical distribution system in good condition
- Effective space allocation
- Adequate personal facilities (restrooms, drinking fountains, washup facilities, etc.)

- Efficient traffic flow
- Adequate functional emergency exists
- Effective alarms and communication systems
- Adequate fire prevention and extinguishing devises
- Acceptable interior color scheme
- Acceptable noise absorption factor
- Adequate maintenance and cleanliness

**Machinery and Equipment**

- Acceptable placement, securing, and clearance
- Clearly marked safety zones
- Adequate nonskid flooring around machines
- Adequate guard devices on all pulleys
- Sharp, secure knives and cutting edges
- Properly maintained and lubricated machines, in good working condition
- Functional, guarded, magnetic-type switches on all major machines
- Properly wired and grounded machines
- Functional hand and portable power tools, in good condition and grounded
- Quality machines adequate to handle the expected work load
- Conspicuously posted safety precautions and rules near each machine
- Guard for all pinch points within 7 feet of the floor

## Other Obstacles

### *Flammables*

Flammables are substances that will burn rapidly when they reach their ignition temperature in the presence of an oxidizer.

Flammables encompass a broad range of substances from solids such as paper and rags, to liquids such as gasoline, to gases such as hydrogen.

Operating employees, supervisory personnel, staff, and passersby must all be constantly cautioned against the unauthorized presence of flammables, especially in designated areas. Poor housekeeping discipline is one of the primary causes of the illegal presence of flammables.

Smokers or other sources of open flames in combination with flammables can become the team that burns buildings down.

### *Hazardous Arrangements of Equipment*

Equipment position is an important consideration for a variety of reasons.

Unplanned equipment arrangement can result in inefficient production and can be a direct cause of lost profits.

Less obvious, but important nonetheless, is the necessity to consider equipment layout to permit the free and unobstructed flow of personnel and materials through a facility.

Poorly arranged equipment has been known to cause accidents that could have been prevented. Lack of sufficient room for free movement by machine operators and mobile equipment, as well as ample space for waste storage, materials awaiting work or movement, and so on, can be the cause of injury to personnel as well as damage to equipment and facilities.

Personnel should be encouraged to identify potential problem areas caused by the hazardous arrangement of equipment.

## PERSONAL PROTECTIVE EQUIPMENT

### Background

In many cases the use of personal protective equipment is a safety procedure of last resort. It is worn when safety has not been designed into a machine or operation, or a hazard designed out of it. In other instances personal protective equipment is required to protect employees from themselves—for example, safety shoes in case someone drops a heavy wrench on his or her toe.

Regardless of whether personal protective equipment is a last resort procedure, however, the employer, as provided under the OSHAct, must fully understand its intended purpose and how this relates to job procedures, and this information must be conveyed to the employees.

OSHAct's Subpart I, "Personal Protective Equipment," is not comprehensive. In most categories the reader is referred to the relevant ANSI standard.

The first section, 1910.132 ("General Requirements"), states that protective equipment—for eyes, face, head, and extremities; protective clothing; respiratory devices; shields and barriers—is considered to be personal protective equipment, and that such items shall be provided, used, and maintained where circumstances warrant.

It also says that where employees provide their own equipment, the employer is responsible for its adequacy, maintenance, and sanitation.

The section further states: "all personal protective equipment shall be of safe design and construction for the work to be performed."

In 1976 there were approximately 2.2 million disabling work injuries in the United States. Of these, about 12,500 were fatal and 80,000 resulted in some permanent disability. The approximate numbers of injuries involving different parts of the body, as estimated from the combined reports of state labor departments, are as follows:

| | |
|---|---|
| Trunk | 600,000 |
| Fingers | 350,000 |
| Legs | 280,000 |

| | |
|---|---|
| Arms | 200,000 |
| General | 180,000 |
| Eyes, feet, hands, head (except eyes) | 130,000 |
| Toes | 70,000 |

It can be assumed that at least some of these injuries could have been avoided if appropriate, clean, well-maintained protective equipment had been worn.

In an effort to increase the likelihood that such equipment will be worn, it is necessary to provide all employees with suitable and timely training regarding its adequacy, maintenance, and sanitation.

Additionally, all employees should be completely familiar with the following matters:

- Company policy on payment for personal protective equipment
- Every area where any type of protective equipment must be worn
- Company policy on violations of rules regarding personal protective equipment
- Company policy on issuing, repair, and replacement of equipment

Employee resistance to the use of personal protective equipment can be significantly reduced through the development of a positive attitude about its use. Motivational training on this subject is easy to present initially, but desired employee attitudes are difficult to maintain over the long haul.

### Planning, Implementing, and Maintaining the Use of Protective Equipment

This section reviews ways to help plan, implement, and maintain personal protective equipment. This can be considered in terms of the following nine phases: (1) need analysis, (2) equipment selection, (3) program communications, (4) training, (5) fitting and adjustment, (6) target date setting, (7) break-in period, (8) enforcement, and (9) follow-through.

The first phase of promoting the use of personal protective equipment is called *need analysis.* Before selecting protective equipment, the hazards or conditions the equipment must protect the employee from must be determined. To accomplish this, questions such as the following must be asked:

- What standards does the law require for this type of work in this type of environment?
- What needs do our accident statistics point to?
- What hazards have we found in our safety and/or health inspections?
- What needs show up in our job analysis and job observations activities?
- Where is the potential for accidents, injuries, illnesses, and damage?
- Which hazards cannot be eliminated or segregated?

The second phase of promoting the use of protective equipment is *equipment selection*. Once a need has been established, proper equipment must be selected. Basic consideration should include the following:

- Conformity to the standards
- Degree of protection provided
- Relative cost
- Ease of use and maintenance
- Relative comfort

The third phase is *program communication*. It is not appropriate to simply announce a protective equipment program, put it into effect, and expect to get immediate cooperation. Employees tend to resist change unless they see it as necessary, comfortable, or reasonable. It is helpful to use various approaches to publicity and promotion to teach employees why the equipment is necessary. Various points can be covered in supervisor's meetings, in safety meetings, by posters, on bulletin boards, in special meetings, and in casual conversation. Gradually, employees will come to expect or to request protective equipment to be used on the job. The main points in program communication are to educate employees in why protective equipment is necessary, and to motivate them to want it and use it.

*Training* is an essential step in making sure protective equipment will be used properly. The employees should learn why the equipment is necessary, when it must be used, who must use it, where it is required, what the benefits are, and how to use it and take care of it. Do not forget that employee turnover will bring new employees into the work area. Therefore you will continually need to train new employees in the use of the protective equipment they will handle.

After the training phase comes the *fitting and adjustment* phase. Unless the protective equipment fits the individual properly, it may not give the necessary protection.

There are many ways to fit or to adjust protective equipment. For example, face masks have straps that hold them snug against the contours of the head and face and prevent leaks; rubberized garments have snaps or ties that can be drawn up snugly, to keep loose and floppy garments from getting caught in machinery. The important point is to make sure the personal protective equipment really fits the employee, or else it will not provide the protection it is intended to give.

The next phase is *target date setting*. After the other phases have been completed, set specific dates for completion of the various phases. For example, all employees shall be fitted with protective equipment before a certain date; all training shall be fitted with protective equipment before a certain date, all employees must wear their protective equipment while in the production area.

After setting the target dates, expect a *break-in period.* There will usually be a period of psychological adjustment whenever a new personal protective program is established. Remember two things:

- You can expect some gripes, grumbles, and problems.
- Appropriate consideration must be given to each individual problems, then strive toward a workable solution.

It might also be wise to post signs that indicate the type of equipment needed. For example, a sign might read "EYE PROTECTION MUST BE WORN IN THIS AREA."

After the break-in phase comes *enforcement.* If all the previous phases were successful, problems in terms of enforcement should be few. In case disciplinary action is required, sound judgment must be used and each case evaluated on an individual basis. Some employers follow the rules below for disciplinary action.

- First offense: an oral warning
- Second offense: a written warning
- Third offense: one week's suspension
- Fourth offense: two week's suspension
- Fifth offense: discharge

These actions may seem harsh, but if employees fail to use protective equipment, they may be exposed to hazards. Do not forget, the employer can be penalized if employees do not use their protection.

The final phase is *follow-through.* Although disciplinary action may sometimes be necessary, positive motivation plays a more effective part in a successful protective equipment program. One type of positive motivation is a proper example set by management. Managers must wear their protective equipment, just as employees are expected to wear theirs.

Using protective equipment as a subject of safety talks and demonstrations can also have a positive motivating effect. Positive motivation can be gained from including personal protection in themes (carefully designed buttons, badges, banners, and brochures) and in contests that can lead to awards for those with exemplary performance in the use of protective equipment. Additional positive motivating forces are safety clubs that give recognition to employees for avoiding serious injury by proper use of their personal protection. A few examples of safety clubs that are active in certain establishments are as follows:

- Turble Club: hard hats
- Scarab Club: safety shoes

- Halfway to Hell Club: safety nets
- Wise Own Club: eye protection

### *Adequacy, Maintenance, and Sanitation*

Before selling safety shoes and supplying safety goggles at a company store, the attendants must be guided by a well-structured program of equipment maintenance, preferably preventive maintenance.

Daily maintenance of different types of equipment might include adjustment of the suspension system on a safety hat; cleaning of goggle lenses, glasses, or spectacles; scraping residue from the sole of a safety shoe; proper adjustment of a facc mask when donning an air-purifying respirator.

Performing these functions should be coupled with periodic inspections for weaknesses or defects in the equipment. How often this type of check is made, of course, depends on the particular type of equipment used. For example, sealed-canister gas masks should be weighed on receipt from the manufacturer, and the weight should be marked indelibly on each canister. Stored units should then be reweighed periodically, and those exceeding a recommended weight should be discarded even though the seal remains unbroken.

Sanitation, as spelled out in OSHAct, is a key part of any operation, and it requires the use of personal protective equipment, not only to eliminate cross-infection among users of the same unit of equipment, but because unsanitary equipment is objectionable to the wearer.

Procedures and facilities that are necessary to sanitize or disinfect equipment can be an integral part of an equipment maintenance program. For example, OSHAct says, "Respirators used routinely shall be inspected during cleaning." Without grime and dirt to hinder an inspection, gauges can be read better, rubber or elastomer parts can be checked for pliability and signs of deterioration, and valves can be checked.

## Eye and Face Protection

When confronted with employee comments such as, "it's too tight," "it pinches my nose," or "these glasses blur my vision," safety professionals may throw up their hands in despair over the plant eye safety program.

To meet the requirements of the OSHAct regulation on eye and face protection, both employer and employee must cooperate: the employer must "make conveniently available a type of protector suitable for the work to be performed," and the employee "shall use such protectors."

To encourage this "use" by the employee, the law goes into detail about how the eyewear should fit and how it should be maintained. Provisions are made for employees who must have their vision corrected to perform their jobs. Phrases such as "reasonably comfortable," "fit snugly," and "easily cleanable" dot the wording of the regulation.

With this legislative weight, the safety professional has a strong selling point for the eyewear program. In addition to an explanation on equipment designed for specific types of hazard protection, a selection and fitting program is necessary. The elaborateness of such a program depends, of course, on the types of protection needed and the size of the workforce. However, informed and trained personnel should be used regardless of the size of the program. Not everyone knows how to use a pair of optical pliers to fit safety glasses, nor can just anyone spot allergic reactions to certain types of metal frames. But information such as "scratched or pitted lenses are injurious to vision, and can alter impact resistance" can be conveyed to all employees in the eyewear program by the safety professional. Self-monitoring preventive maintenance is much simpler to promote when employees are well informed of their responsibilities in a protective equipment program.

"Eye and Face Protection," CFR 1910.133, establishes the general requirements that no unprotected person shall knowingly be subjected to a hazardous condition and that "suitable eye protectors" shall be provided in the presence of flying objects, glare, liquids, injurious radiation, or any combination of these.

Seven minimum requirements are set, as follows:

- Adequate protection against the particular hazards the device was designed for
- Reasonably comfortable when worn under the designed conditions
- Snug fit and no undue interference with the wearer's movements
- Durability
- Capable of being disinfected
- Easily cleanable
- Kept clean and in good repair

The section also specifies that persons who need optical correction and eye protection shall wear either corrective eye protectors or goggles that can be worn over the protective lens. Paragraph 1910.133 also says that use will be in compliance with ANSI Z87.1-1978, "Occupational and Educational Eye and Face Protection." Other requirements concern the manufacturer.

## Hand Protection

The OSHAct subpart on personal protective equipment has only one brief regulation referring to hand protection. Other sections of the law touch on hand safety with reference to tools and machine guarding, but Section 1910.137 requires rubber insulating gloves for electrical workers.

More standards are needed to cover the myriad industrial procedures in which hands require the protection of gloves. Items such as finger cots, head- and abrasion-resistant hand pads, and barrier creams, not to mention leather, cloth, asbestos, rubber, and neoprene gloves, are all forms of hand protection. Industrial operations meriting their use will eventually be spelled out. For example, nagging (and sometimes painful) itch of industrial dermatitis can be effectively counteracted with barrier creams and appropriate gloves.

Preemployment physicals are not meant to analyze a person's health in depth, but a predisposition to allergic reactions and other diseases can be recorded. Based on such information, appropriate hand protection can be assigned to the worker who may react to any chemicals or solutions issued on the job. (Visual acuity, lifting restriction, and hearing problems are other areas checked during a preemployment physical that can also determine what other types of personal protective equipment are assigned to a worker.)

There is no all-purpose form of hand protection. The material used for gloves depends largely on what is being handled.

- For most light work, a canvas or other cloth glove is both satisfactory and inexpensive
- For rough or abrasive material, leather or leather reinforced with metal stitching is required. Leather reinforced by metal stitching or metal mesh also provides good protection from edged tools.
- There are many plastic and plastic-coated gloves available, and they are designed to give protection from a variety of hazards. Some surpass leather in wearing ability. Others have granules or rough materials incorporated in the plastic for better gripping ability. Some are disposable.
- Asbestos gloves and combination gloves are available for handling hot materials.
- Metal mesh, terry cloth, and a variety of other glove constructions are available in a variety of styles to match hand protection with the particular hazard.
- Finger stalls of various materials, in combinations of one or more fingers, are available for situations that do not require a complete glove.
- Gloves coated with rubber, synthetic elastomers, polyvinyl chloride, or other synthetics offer protection against all types of petroleum products, caustic soda, tannic acid, and hydrochloric acid. They are also recommended in the handling of sulfuric acid. These gloves are available in varying degrees of strength to meet individual conditions.

Barrier creams, if properly used and reapplied frequently, provide limited protection against irritants to the hands and arms. There are four common types based on function—soap base, solvent-repellent, water repellent, and special types. No one cream is effective against all irritants. The need for repeated washing to remove the barrier cream is one principal benefit from its use.

### General Body Protection

OSHAct regulations stipulate that protective clothing shall be provided, used, and maintained. Yet no specific standards are listed for clothing that gives general body protection, such as jackets, aprons, trousers, lab coats, and lifelines.

With the advent of disposable clothing, a number of manufacturers now offer product lines suitable for use in clean-room operations. Some garments have antistatic and flame-retardant qualities that allow for repeated launderings. Other types of apparel are for one-time use only.

The design of garments such as leggings and sleeves that augment other types of protective equipment—gloves, respirators, boots—should facilitate wearing the basic protective devices or equipment. Sleeves should not be so tight fitting at the wrist that gloves cannot be worn comfortably and properly. Leggings should fit close to the leg so as not to catch in moving machinery.

Lifelines, which are worn by employees who work above grade, should be inspected at regular intervals, and all general body protective clothing should be chosen so it can be put on and removed quickly.

Life vests should be worn by those who work over water.

### Respiratory Protection

As a part of a human body increases in importance to sustaining life, the OSHAct regulations governing its protection increase in number. With a preliminary reference to the primary objective of "preventing atmospheric contamination," the law notes that such engineering controls are not always feasible, and there follows a series of explicit regulations on the protection of the respiratory system.

Again stipulating cooperation in the form of "employer shall provide" and "employee shall use," the law requires the establishment of written operating procedures as a part of a minimal acceptable program.

To ensure adequate protection for the wearer of respiratory equipment, the person must be physically capable of performing the job functions while so protected. Periodic reviews are to be made of the medical status of the respirator user.

Once a worker has a clean bill of health to do a job while wearing a respirator, the OSHAct requires that he or she be trained in the use of the equipment and in it limitations. Here the safety professional must choose carefully the way to explain operation of the equipment. It is necessary to refrain from instilling fear of the hazard in the worker without conveying a false sense of security. By pointing out to the employee that the employer must supplement the equipment's use with surveillance and inspections of work area conditions, the safety professional can reinforce a positive attitude about the hazards of a job function that requires respiratory protection.

In addition to a number of regulations on the mechanical parts and air supplies for respirators, the law notes that "possible emergency and routine uses of respirators should be anticipated and planned for." The buddy system is required for areas where a wearer could be overcome by a toxic or oxygen-deficient atmosphere if the respirator failed. Communications (visual, voice, or signal line) are required between both or all individuals present.

In some plants where respiratory hazards are not part of daily operations, respirators are supplied for emergency use. The OSHAct requires that records be kept of equipment inspections, and stipulated maintenance procedures must be followed after each use.

"Respiratory Protection," CFR 1910.134, states that in control of occupational diseases caused by breathing air contaminated with harmful dusts, fogs, fumes, mists, gasses, smokes, sprays, or vapors, the primary objective is to prevent atmospheric contamination. This is to be accomplished as far as feasible by accepted engineering control measures (e.g., enclosure or confinement of the operation, general and local ventilation, and substitution of less toxic materials). When effective engineering controls are not feasible, or while they are being instituted, appropriate respirators are to be used pursuant to the following requirements:

- Respirators shall be provided by the employer when such equipment is necessary to protect the health of the employee.
- The employer shall provide the respirators that are available and suitable for the purpose intended.
- The employer shall be responsible for the establishment and maintenance of a respiratory protective program.
- The employee shall use the provided respiratory protection in accordance with instructions and training received.

The requirements for a minimal program are as follows:

- Written standard operating procedures governing the selection and use of respirators shall be established.
- Respirators shall be selected on the basis of hazards to which the worker is exposed.
- The user shall be instructed and trained in the proper use of respirators and their limitations.
- Where practicable, the respirators should be assigned to individual workers for their exclusive use.
- Respirators shall be regularly cleaned and disinfected. Those issued for the exclusive use of one worker should be cleaned after each day's use, or more often if necessary. Those used by more than one worker shall be thoroughly cleaned and disinfected after each use.
- Respirators shall be stored in a convenient, clean, and sanitary location.
- Respirators used routinely shall be inspected during cleaning. Worn or deteriorated parts shall be replaced. Respirators for emergency use, such as self-contained devices, shall be thoroughly inspected at least once a month and after each use.
- Appropriate surveillance of work area conditions and degree of employee exposure or stress shall be maintained.

- There shall be regular inspection and evaluation to determine the continued effectiveness of the program.
- Persons should not be assigned to tasks requiring use of respirators unless it has been determined that they are physically able to perform the work and use the equipment. The local physician shall determine what health and physical conditions are pertinent. The respirator user's medical status should be reviewed periodically (e.g., annually).
- Approved or accepted respirators shall be used when they are available. The respirator furnished shall provide adequate respiratory protection against the particular hazard for which it is designed in accordance with standards established by competent authorities.

Proper selection of respirators shall be made according to the guidance of American National Standard "Practices for Respiratory Protection" (Z88.2-1969).

### Head Protection

The OSHAct requires protective helmets to meet the requirements of ANSI Z89.1-1966 for impact and penetration resistance from falling and flying objects, but it is up to the safety professional and appropriate supervisors to make sure that the equipment is worn in the recommended manner.

To counter any deliberate employee mutilation (or misuse) of a safety helmet or bump cap, as frequently happens, the cost of replacement can be borne by the employee. The cutting or removal of suspension straps does not make a 15½ ounce helmet (maximum weight as defined by ANSI Z89.1) any more comfortable, and the effectiveness of the protection is diminished. Supervisors should be instructed by the safety professional to be thorough in their inspection for this type of equipment damage.

One type of head protection not provided for in the OSHAct is the protection of hair, flame-resistant net-type caps (where necessary), disposable nets, bandeaux, and other types of lightweight caps are available for both male and female industrial workers. These types of protection should be worn by any employee who choose to wear their hair in longer styles and whose hair is subject to entanglement in moving machinery or to ignition from sparks.

"Occupational Head Protection," CFR 1910.135, is very brief and merely refers the reader to ANSI Z89.1-1969, "Safety Requirements for Industrial Head Protection."

The standard is intended primarily to guide manufacturers of hard hats in the construction and testing of protective headgear, but it might be well to know the characteristics of hats made to this standard as an aid in selection.

The standard divides hats into two types:

- Type A. A helmet with a full brim at least 1½ inches wide.
- Type 2. A cap with no brim, but with a peak extending forward from the crown over the eyes only.

Each type is divided into four classes:

- Class A. Provides limited voltage protection.
- Class B. Provides greater voltage protection.
- Class C. Provides no voltage protection.
- Class D. Provides limited voltage protection and is designed for fire fighting—it exists only in the form of Type I.

All classes must provide the same protection from impact—that is, the helmet and its suspension may not transmit an average force of more than 850 foot-pounds, and no individual specimen may transmit more than 1000 foot-pounds when tested under prescribed standard conditions at temperatures of both 0 and 120°F.

Penetration allowable is slightly more for Class C helmets than for other classes.

Classes A and D, when tested under standard conditions at 2000 volts AC for 1 minute, must allow leakage of no more than 3 milliamperes. Class B (under ANSI Z2.1, which is still in effect for this class) must take 15,000 volts AC for 1 minute with leakage of no more than 8 milliamperes. Class C helmets need offer no insulation from electric current.

Classes A and B must be a slow-burning material, and Class D must be of material that does not support combustion. Class C helmets are not tested for flammability, since they are made of metal.

All classes must restrict absorption of water under standard test conditions to no more than 5% of the weight of the helmet.

Standards call for the prominent labeling of all helmets with the letter indicating their class.

It is important to remember that the protection offered by a safety hat or cap depends on both the helmet shell and the suspension that separates it from the head. A defective or badly adjusted suspension permits the transmission of excessive force from the hard shell to the head.

Suspensions are not interchangeable between Class B helmets and other classes. Class B suspensions must include no metal parts and may not be supported by fittings that require the drilling of holes through the helmet.

Size is of critical importance in suspensions. Some companies use only nonadjustable suspensions, to ensure against misadjustment by the worker. But whether the adjustable or nonadjustable type is chosen, the suspension in use must be right for the individual who wears it.

The protective quality of a hat or cap can be ensured by purchasing it only from a reliable manufacturer or supplier and by careful inspection before it is put in use.

Finally, the best hat or cap in the world can deteriorate in use both from normal wear and tear and from lack of proper care and maintenance. Surveys have shown, for example, that many helmet shells are used after they have developed cracks, pits, or dents, which diminish their protective quality. Suspensions can become torn, frayed or filthy.

Although OSHAct regulations make the employer responsible for proper hard hat maintenance, it does not define "proper maintenance." The following is general industry practice on hard hat maintenance:

- Good maintenance begins with the worker who uses the hat or cap. He or she should be provided with a secure, clean place to store headgear, where it is protected against blows, dirt, temperature extremes, and harmful chemicals. This is especially important when workers carry the headgear in trucks or cars. A safety cap or hat bouncing around with other equipment on the bed of a truck is not likely to remain in good condition long. A common place for carrying caps and hats in a car is the shelf behind the back seat. Here it is subject to damage and to excessive heat when the car is parked on a sunny day. Here, also it is likely to be turned into a missile if the car stops suddenly.
- Every part of each safety hat and cap should be visually inspected daily for signs of wear, damage, cracks, dents, pits, or other flaws. Inspection should include the helmet shell, the suspension, the sweat band, and the accessories. Any headgear with defects that reduce its protective quality should be removed from service and replaced. Correction of the fault may be possible, but the headgear should not be used until correction is made.
- Cleanliness is of major importance, both to protect the worker against infection and irritation and to encourage the regular use of the protection by keeping it attractive. Companies should secure from the supplier or manufacturer instructions for cleaning both shell and suspension by means that are effective but do not damage the material. The responsibility for cleaning and sterilizing cannot be safely left to the individual employee.
- When hats or caps are shifted from one employee to another, the helmet should be carefully cleaned and the suspension sterilized or replaced with a new one.
- Sometimes it seems desirable to either the employer or the employee to paint the headgear. Check before allowing this, because some paints are incompatible with some plastic shell materials. It is usually better to purchase shells of colored plastic than to paint the desired color on the shell.
- Headgear can protect only if it is worn. This means that the employer must plan a program to gain employee acceptance of the need for safety headgear and to ensure its regular and continued use.

The need for head protection should be carefully and clearly explained. The kinds of hazard that exist on the job and from which headgear offers protection should be presented clearly and forcefully. The effectiveness of safety hats and caps in preventing injuries should be made clear, but always with the honest acknowledgment that safety headgear, like any type of personal protective equipment, is a last line of defense when engineering and work procedures fail to prevent dangerous events.

It should be stressed that hats and caps protect against more than falling objects. Workers should understand that safety headgear may prevent injuries when the head strikes a fixed object, when it is struck by a moving but not falling object (crane hook, took, etc.), when contact is made with electrical conductors, and when the worker is exposed to splashes of hot material or harmful chemicals.

Emphasis on these points will, in many situations, explain the need for a particular type of safety headgear. The need for a brimmed hat rather than a peaked cap, for example, may be made clear by stressing the likelihood of impacts other than those caused by objects falling vertically. The nature of the job will determine the degree of electrical protection needed, and this point should be thoroughly explained to the worker.

### Hearing Protection

Although the OSHAct regulations set maximum permissible noise levels and exposures, and explain corrective procedures under Subpart G ("Occupational Health and Environment Control"), Section 1910.95, no mention is made in Subpart I ("Personal Protective Equipment") of hearing protection to be worn by the employee.

The Walsh-Healey Act and the references in the OSHAct are supplemented, for better understanding, by the Department of Labor's *Guidelines to Occupational Noise Exposure.*

Both Walsh–Healey and the OSHAct see engineering controls as the first line of defense against noise-induced hearing problems. However, this method of resolving the problem of noise exposure is not always feasible. Hence the need for personal hearing protection equipment, although the government does not term this a permanent solution to the problem.

Beginning with the simple wax-impregnated cotton that the employee shapes and inserts in the ears, the selection of equipment continues through a diverse range, which include cuffs with features for coupling other types of personal protection with ear protection.

Selection of the most appropriate model of equipment depends on the type of noise problem in the employee's work area, the job operation to be performed, and, to a certain degree, the individual's personal preference of equipment.

The preferred way to select the correct ear protection involves preemployment physicals (to determine any problems prior to on-the-job exposure) and the recommendation of an industrial hygienist who has surveyed the work area. Furthermore, only hearing protectors tested in accordance with the ANSI Z24.22-1957, "Method for Measurement of the Real Ear Attenuation of Ear Protectors at Threshold," are acceptable for use according to the Department of Labor.

The seven-part "Directory of Noise Control Products," published in 1971 by National Safety News, includes a section ("Personal Protection," National Safety News reprint stock no. 111.17-61) detailing models of ear protection as follows:

> Inserts. Formed by user, premolded, and custom molded to block external ear canal from noise.
>
> Superaural. Two disks covering ear canal openings and suspended from a headband.
>
> Circumaural. Cup or dome-shaped devices placed over the ear, including the lobe, to reduce sound level before it reaches external auditory canal.

Whatever model of protection is selected, performance and comfort must be considered in addition to the application of the equipment to other job functions.

Educational and promotional programs on personal hearing protection are requirements for implementation of the Walsh–Healey Act and for an "effective hearing conservation program" as specified under the OSHAct, CFR 1910.95 paragraph (b)(3). It should be noted that standards are needed to establish minimum requirements for this type of personal protective equipment.

Regardless of the type of ear protector decided on, its attenuation, as stated by the manufacturer, must be sufficient to reduce the noise level in the worker's ear to the level and for the duration prescribed in Table 7-1. The manufacturer's stated values are determined under ideal conditions; therefore, as a precaution, it is wise to assume that the attenuation actually attained in use in the shop will be at least 5 decibels (dB) less than the stated value. Only ear protectors that have been tested in accordance with American National Standard Z24.22-1957, "Method for Measurement of the Real Ear Attenuation of Ear Protectors at Threshold," are acceptable to the Department of Labor.

The Department strongly recommends that any employee who is exposed to high sound levels and requests ear protection be provided with it, even if the duration of exposure is within the limits prescribed by Table 7-1 and Figure 7-1.

When engineering and administrative controls fail to bring noise levels or duration of exposure to permissible levels, the use of personal protective equipment is required, as stated in the second sentence of paragraph b, Section 50-204.10 (Walsh–Healey Act) and CFR 1910.195:

> If such controls fail to reduce sound levels within the levels of the table, personal protective equipment shall be provided and used to reduce sound levels within the levels of the table.

The use of personal protective equipment is considered by the Department to be an interim measure while engineering and administrative controls are being perfected. In very few cases will the use of this equipment be acceptable as a permanent solution to noise problems.

Some methods of control, such as providing an isolation booth for operators, or conducting noisy operations when few employees are in the plant, may require use of personal protective equipment by the operator when he or she must emerge from the booth to make adjustments, or by the few employees who carry on the noisy operation.

In addition, the regulations require that personal protective equipment be provided and used. How it is used is up to the employer. The Department recommends, however, that an educational and promotional program precede initiation of required use of such equipment, and continue as long as necessary to achieve 100% acceptance by employees. In the absence of an observable high proportion of use, the Department would consider the lack of a training and promotional program to constitute a violation of the regulation.

## Foot Protection

Like Section 1910.135, Section 1910.136, "Occupational Foot Protection," does nothing but refer the reader to a standard—in this case ANSI Z41.1-1967, "Standard for Men's Safety-Toe Footwear."

This standard outlines the specifications, performance requirements, and methods of tests for safety footwear designed primarily to protect the toe.

It divides shoes into three classes:

> Class 75. This shoe must provide 16/32-inch clearance after 75 foot-pounds of impact, and 16/32-inch clearance after 2500 pounds compression.
>
> Class 50. This shoe must provide 16/32-inch clearance after 50 foot-pounds of impact and 1750 pounds compression.
>
> Class 30. This shoe must provide 16/32-inch clearance after 30 foot-pounds of impact and 1000 pounds compression.

The standard also provides that all shoes be clearly marked to indicate the class to which they belong.

Although not stated, it is implied that Class 75 is for heavy hazard exposure, Class 50 for medium exposure, and Class 30 for light exposure.

When selecting shoes, look for the identification of the class on the inside quarter, the shank of the outsole, shank of the insole, or tongue. The class number will be enclosed in a border.

The impact test for Class 75 calls for a 50-pound weight to be dropped from a height of 18 inches with at least ½-inch clearance remaining between the toe cap and the insole of the shoe.

The compression test calls for applying a load, at the rate of 50 pounds per second, after the first 500 pounds of load, to the highest point of the toebox. The clearance between the toe cap and the insole must be at least ½-inch.

Here again, OSHAct regulations make the employer responsible for maintenance but does not specify what constitutes proper maintenance.

In the case of steel-tipped shoes and other safety footwear, maintenance will be largely up to the individual.

Some steps management can take to ensure compliance are as follows:

- Inform electricians with special shoes to avoid the use of nails in repairing.
- Make periodic tests on conductive shoes to make sure the maximum allowable resistance (450,000 ohms) is not exceeded.
- Carefully disinfect the boots after each shift when different shifts use the same boots.

### Protection from Electrical Devices

Section 1910.137, "Electrical Protective Devices," refers the readers to existing standards, namely ANSI J6.1, J6.2. J6.6, and J6.7, which cover rubber insulating gloves, matting, blankets, hoods, line hose, and sleeves. This section deals only with the items that are personal protective equipment—rubber insulating gloves and rubber insulating sleeves.

The ANSI standard covering rubber insulating gloves, J6.6, divides them into five classes:

Class 0 with a minimum breakdown voltage of 6000 rms volts (AC) and 35,000 average volts (DC)

Class 1 with a minimum breakdown voltage of 20,000 rms volts (AC) and 60,000 average volts (DC)

Class 2 with a minimum breakdown voltage of 25,000 rms volts (AC) and 70,000 average volts (DC)

Class 3 with a minimum breakdown voltage of 30,000 rms volts (AC) and 80,000 average volts (DC)

Class 4 with a minimum breakdown voltage of 45,000 rms volts (AC) and 100,000 average volts (DC)

Maintenance of rubber gloves and sleeves consists mainly of washing. Mild soap flakes or a household detergent dissolved in warm water (about 90 to 100°F) can be used. Special compounds are also available commercially. Gloves and sleeves are then washed for about 15 minutes in a tumbler or agitator-type washer.

After washing the garments are rinsed in clear water and dried thoroughly. If a home-type dryer is used, the maximum temperature should be 150°F. A temperature of 120 to 130°F is recommended. But ozone-producing light bulbs should be removed from the dryer.

After inspection and retesting, the protective gear can be lightly dusted with soapstone or commercial talc and placed in stock.

All sleeves and gloves issued from storerooms or tool rooms should be visually checked for injury or damage that may have occurred in transit or storage. Particular attention should be given to equipment returning from special jobs (e.g., storm emergency).

Equipment showing tears, scratches, cuts, cracks, burns, or having a "dead feeling" should be tagged "defective" and taken out of service. The proper individual should then decide whether the item should be returned to laboratory for retest.

Rubber gloves can be inspected by filling them with air or water and seeing if there are any leaks.

For rubber sleeves, stretching or rolling the rubber between the fingers will aid in revealing defects that normally cannot be seen if the sleeve is lying flat.

As a general rule such equipment should not be stored when wet or dirty or in a manner that will distort or change a device from its natural shape. Rubber will take a permanent "set" if distorted from its original shape for a long time.

## Protection from Environmental Hazards

Thus far OSHA has not promulgated any regulations dealing with control of two other types of industrial environmental hazards—temperature extremes and atmospheric pressure extremes. Responsible personnel should, however, be aware of the problems associated with these hazards.

### *Temperature Extremes*

Extremely cold atmospheres are routine in meat processing, and they can cause problems in any industry that uses outdoor teems in cold climates.

The basic precaution against cold is proper clothing. It should be warm, allow perspiration, and not restrict circulation. Preferably there should be several layers of clothing. This provides layers of trapped air and air is a good insulator.

At below-zero temperatures, workers should be instructed to wear gloves and avoid touching metal. Whenever possible, wood-handled tools should be provided.

If workers must work out of doors in very cold weather, they should be informed about the precautions and treatments associated with frostbite.

Whether working outdoors or in a deep-freeze, they should be cautioned to be careful while walking on frost and ice.

There are three approaches to contending with extreme heat—shielding, acclimatization, and personal protection.

Reflective screens, heat exchange screens, absorbing screens, and aluminized cloth curtains can be used to shield workers from radiant heat. In addition, a booth can be provided to shield personnel.

After about 10 days of relatively short daily exposures to a certain amount of heat, a person becomes acclimated. He or she is then able to perform a hot job with considerably less physical stress than an unacclimatized person would experience.

The beneficial effects of acclimatization are easily attained over a weekend, but they will be lost if there is no exposure to heat for a period of about two weeks.

Personal protection from extreme heat includes the vortex suit (which can also protect against cold) and reflective clothing.

### *Atmospheric Pressure*

Like temperature, atmospheric pressure must be dealt with at high and low extremes.

The main hazard of operations performed under high atmospheric pressure is decompression sickness or the "bends." Some of the controls of this condition include a properly equipped and manned decompressive chamber, periodic medical checkups for workers, screening out of alcoholics, and use of an identification badge stating that the individual is a compressed air worker. (Because the symptoms of the bends resemble those of drunkenness, victims are often taken to jail rather than to a decompression chamber.)

Low atmospheric pressure is usually associated with high mountain elevations, but is also fund in some aerospace situations. The usual approach is pressurization or supplied-air masks.

A vortex suit is a generic classification of a type of clothing that provides total body coverage (including head, hands, and feet) and that uses an external supply of circulating air that provides a constant ambient temperature for the wearer (e.g., astronaut's space suit).

Although these and other areas normally considered under the heading of industrial hygiene are not now covered by specific OSHAct regulations, subsequent directives may be issued that will spell out exacting procedures. Prudence requires employers to take advantage of whatever knowledge is available in these areas and to incorporate it in their industrial hygiene programs.

## SUMMARY

The level of detail provided for the above information concerning OSHA rights and responsibilities, housekeeping, and personal protective equipment will obviously vary from company to company and job to job, but the essential elements of these three topics should be a cornerstone of safety and health for new employee training programs.

## CHAPTER 7

# EMERGENCY ACTIONS AND REPORTING

### INTRODUCTION

Even though it is straightforward and simple in terms of subject matter complexity, emergency training should be an essential element in all safety and health training programs.

This should be some of the first training presented to new employees and new supervisors. Ideally, this training should be accomplished the first day on the job.

### EMERGENCY ACTIONS

Emergency actions refers to those things an individual should do, in sequence, in an emergency situation. Emergency action training is very important and, to be effective, must be periodically reinforced with refresher training and practice.

Studies have indicated that in emergency situations, human error rates are very high (20% and up) regardless of training. Thus, it is important that the tasks that individuals are to perform in an emergency situation be very simple and straightforward and that training be intense (Figure 7.1).

Your first responsibility in an emergency is to protect yourself and avoid becoming a victim. Then seek assistance by reporting the information in accordance with your training and your organizations emergency plan. Finally,

Emergency Response Actions

- Escape!!!
  Protect yourself
- Report!!!
  Know how, what, and to whom
- Assist
  Within capabilities and
  without self endangerment

**FIGURE 7.1** Emergency response actions.

within your capabilities, try to rescue or assist victims and/or stop or slow the damage without endangering yourself.

### All Employees

Even though there may be many tasks required of persons in certain occupations (airline pilots, nuclear power plant operators, ship captains, etc.), the primary tasks for most employees are to (1) get our of harms way (ESCAPE!), (2) call for help, and (3) within capabilities, if the situation permits, take actions to ameliorate the situation.

#### *Getting out of Dodge!!!*

Emergency training must, at a minimum, explain in detail how to evacuate the workplace. This training must include the types of alarms or signals or other indications of an emergency or impending or potential emergency. It must also include evacuation routes, assembly areas, and assembly area actions. It may include specific tasks to be taken during the evacuation (assist handicapped employees, turn off or shut down equipment of systems, close doors, seal buildings, etc.).

If the evacuation requires more than simply walking out of a building or walking to a certain part of the worksite, it may be necessary to develop loading plans for vehicles (or other modes of transportation).

All employees must clearly understand the various types of alarms and signals and other means of receiving notification of an emergency. Ideally, emergency alarms will include audible and visual signals, ensuring that both hearing- and sight-impaired employees are notified and providing redundant alarms for other employees.

Even though emergency plans should be as simple as possible, they should include all plausible situations, including acts of nature (earthquakes, floods, tornados, hurricanes, etc.).

It should also be noted that "getting out of harms way" does not always involve evacuation. For certain types of emergencies the appropriate action may be to stay indoors or take cover indoors. Alarm systems must provide clear guidance to prompt persons to take the appropriate action. Complex signals tend to be confusing. If multiple types of alarms are necessary, the emergency action training must be comprehensive and must be periodically reinforced with practice drills and/or refresher training.

If emergency evacuation requires the use of fire escapes, fire slides, or other not commonly used routes or means of egress, the training should include the actual use of these routes/devices.

Assembly areas and alternates should be designated and there should be some simple, clear method of ensuring that everyone knows where to go and what to do after evacuating.

In most cases, at a minimum, there should be a simple, systematic method of accounting for people after the evacuation. Employees should not (unless called for in the emergency plan) leave the assembly area without notifying supervision. Many long, stressful, costly, and sometimes dangerous searches have been made trying to locate and account for employees who are asleep at home or "recovering" in a local bar.

### *Calling for Help!!!*

After getting out of harms way, or if conditions permit and danger is known not to be imminent, a second concern should be to call for help.

The emergency phone number (and/or radio call sign) should be well advertised and ideally located on all telephones and other prominent places in the workplace. For construction workers, the inside brim of the hard hat is a good location. The best systems have only one number to call whatever the nature of the emergency.

Employees need to know the correct information to provide as well as the telephone number or radio call sign to use. See Accident Reporting below.

Training should also include specific instructions to clear radio nets and to stay off the telephone (except for emergency-related communications) when emergencies are in progress. Employees should also be trained to instruct their families to refrain from calling or coming to the worksite during emergencies.

### *Amelioration*

Finally, if and only if there is no imminent danger, employees may take a number of actions to minimize the effects of a possible disaster or other emergency situation. For example, emergency planning may call for cutting off certain equipment or shutting down certain systems.

Evacuation plans may encourage that certain valuable records, computer disks, or other critical items be locked away or carried out. Training should include prioritizing items to be safeguarded or evacuated. Under no circumstances should individuals "go back" after items forgotten or left behind.

Emergency planners should *not* rely heavily on persons completing complex tasks reliably or locating and evacuating hard to find or hard to access items. Human reliability in emergency situations is very low and emphasis should be on preservation of life.

While there are certainly times when it is appropriate for trained employees to fight small fires and "handle" or control small emergencies, a primary reason for having fire extinguishers and other emergency equipment in the workplace is to aid personnel in successfully evacuating the area.

Training should emphasize personal safety. In most cases it is best to leave the fire fighting or other emergency responses to those trained and equipped to do the job.

### Supervisors

In addition to generally taking the same actions as other employees, supervisors generally have additional responsibilities in emergency situations.

Many programs instruct individual employees to report emergencies directly to their supervisor. Therefore, the supervisor must make decisions and take actions to evaluate the location, summons emergency response personnel, and/or take other actions to control the emergency.

In any event, each supervisor must ensure that his or her personnel know and understand emergency procedures, signals, and evacuation routes. Additionally, in case of emergency, the supervisor must oversee the evacuation and/or other emergency actions.

When the situation permits, the supervisor should also preserve the accident site for accident investigators. This involves roping or taping off the area, locking doors or gates, or other controls to keep the accident site from being disturbed.

Also, when the situation permits, the supervisor should have persons who were present at the time of the accident, but are now safe in an assembly area or other location, sit down and write out statements of where they were and what they saw or heard, i.e., their version of the accident to include events leading up to the accident, conditions at the time of the accident, and any other relevant information that may aid in the accident investigation.

This is important for three reasons. First, these preliminary statements can provide some of the most accurate and reliable information available to investigators. Second, these preliminary statements aid the investigators in determining the order in which to call witnesses and in validating witness information. Third, after a serious accident, especially one in which someone was killed or seriously injured, witnesses to the accident may be very upset and nervous and generally feel a need to "do something." This exercise provides these individuals with something meaningful to do and aids in calming nerves.

Most supervisors will not preserve the accident site, have personnel write out preliminary statements, and take other appropriate actions unless trained to do so. This training should be reinforced with practical exercises and emergency drills.

## EMERGENCY REPORTING

One of the essential elements in any new employee orientation, supervisors training program, periodic refresher training program, and safety meeting program should be how to report emergencies. Again it is important to remember that human reliability is low and human error rates are very high in emergency situations, therefore the reporting procedures should be simple and should be frequently reinforced and practiced (Figure 7.2).

### All Employees

Basically every employee must know what to report, to whom to report, how to report (which communication channels to use), and what information to report.

#### *What to Report*

All employees should have a clear understanding of the types of emergencies or other situations to report. These will normally include the obvious (serious injury accidents, major property losses, processes out of control, etc.) but may

**Emergency Reporting**

- What to Report
  Accidents, incidents, near misses, spills, releases, etc. (per Emergency Plan)
- To Whom
  "911", supervisor...(per Emergency Plan)
- How
  Telephone, radio, in person (per Emergency Plan)

**FIGURE 7.2** Emergency reporting.

also include first-aid injuries and other early indications that things may be going wrong (smells of smoke, abnormal instrument readings, items out of place, unidentified items or substances in the workplace, spills of leaks, near misses, etc.). As a general rule, most companies want employees to report any condition or event that is abnormal or has the potential for causing an accident or other unwanted event. The rule of thumb generally should be, "if in doubt, report." Generally, overreporting is not seen as a major concern. Underreporting can be a very major and costly (both to the individual who fails to report and to the organization) problem.

Many people, however, tend to be embarrassed if they report something that turns out to be a "false alarm." It is important that employees be commended for their efforts even in the case of reported incidents that turn out to be insignificant or false alarms.

### *To Whom to Report*

Usually, there are two options for reporting. The first option is to report to the direct supervisor. The second option is to report directly to the appropriate emergency response organization (ideally there is only one emergency telephone number to call which automatically contacts the fire, medical response, security or police, hazmat team, or other responders, as appropriate). This decision may be spelled out in company policy (i.e., always report to the supervisor) or the decision may be left to the discretion of the employee. In the latter case, company guidance should provide criteria for making the decision.

Generally, if the supervisor is readily available (exact location is known and can be reached in less than 10 or 15 seconds), it is desirable to report to the supervisor. *Note*: In emergency situations, decisions must generally be made very rapidly and actions must be taken in a matter of seconds.

If the supervisor is not readily available, it is recommended that all employees have the authority (and responsibility) to contact a higher level of supervision or management and/or emergency response personnel directly. In such a situation, higher level supervision or management should be contacted first only if they are readily available (accessible within less than 15 seconds) and are in a position to rapidly assess the situation.

In any situation where the need for rapid response is obvious (serious injury, explosion, fire, etc.) or if supervision or management is not readily available, it is recommended that every employee has the authority and the responsibility to contact emergency response personnel directly, to activate emergency alarms, and to take other appropriate actions (shut down equipment, evacuate, etc.).

Again, training must emphasize that individuals will not be criticized or disciplined for taking action in what is or is perceived to be emergency or "problem" situations as long as the action was taken in good faith and was not in direct violation of well-communicated and well-understood requirements.

### How to Report

All employees must clearly understand the communication channels available for emergency reporting. An obvious choice where reporting is done to an on-site supervisor is face-to-face communications. For many large organizations or operations where work may be spread over large geographic areas, communications with supervision or emergency response organization must be done by telephone or radio.

In situations where telephones must be used for emergency reporting, it is imperative that all personnel know the location of telephones in the workplace and know the appropriate number(s) to call. Ideally, a single emergency number such as "911" is available and is widely known by the workforce. In all cases, especially if a single number is not available, emergency numbers should be displayed on or next to every workplace telephone and/or on a pocket card issued (and reissued as necessary) to every employee. These numbers should also be displayed prominently in the workplace and should be emphasized in training (should be a test item). An even better system may be to have direct lines and special telephones for emergency reporting (i.e., "red" phones that automatically connect with emergency response coordinators when the receiver is lifted).

For construction and maintenance crews, aircraft, truckers, and other mobile operations or operations in remote areas, the radio may be the only (or best) means of emergency communication. In these cases, it is essential that all personnel be trained in the proper use of the radio, to include the channels and callsigns required for emergency reporting. This information should also be clearly displayed on or near every radio and/or issued on pocket cards and should be displayed throughout the worksite. Training should also include instructions to "clear the net" when emergencies are in progress.

Other means of emergency reporting could be alarms or "panic buttons" located throughout the worksite that automatically notify emergency response personnel and trigger evacuation alarms or prerecorded instructions.

### What Information to Report

In addition to ensuring that all employees know what to report, to whom, and by what means, it is also important that all employees be trained to provide clear, concise, relevant information when reporting emergencies. Generally, the key items of information include the nature of the problem (injuries, fire, explosion, spill, process or equipment out of control, etc.) and the exact location of the problem (to include, if necessary, detailed instructions on how to get to the location). The person reporting should also state his or her name, job title, company, and the telephone number from which they are calling or where they can be reached (or the radio net and callsign).

In most cases, the person receiving the emergency information will be trained to solicit additional information such as a more detailed description of the problem (how many people are injured, what is the nature of their injuries, what hazardous materials are involved, is the facility evacuated, where are the people who were evacuated, is the electrical power still on, etc.).

Reporting personnel should be trained to provide initial information and (*situation permitting*) stay on the telephone or radio until released by the person to whom the report is being made.

### Supervisors

The reporting responsibilities of supervisors are basically the same as for other employees. When notified of an emergency, either by first hand knowledge or by another employee, the supervisor may also have to make a decision as to immediately notify emergency response personnel or to further evaluate the situation and/or to pass the information to the next level of supervision or management.

Since time is usually critical, it is recommended that generally the first supervisor becoming aware of an emergency should immediately report directly to emergency response personnel (depending on circumstances, this may be before or after ordering an evacuation of the site and positioning himself or herself out of danger) and then, time and circumstances permitting, report the problem to upper management.

## SUMMARY

This chapter has probably overused the word "generally." While it is important that emergency actions and reporting be an important part of any safety and health training effort, it is also important that this information be specific and tailored to specific work locations, operations, and processes and be configured to "fit" the communications systems available and the specific emergency response organizations available, taking into consideration their capabilities and limitations.

# CHAPTER 8

# HAZARD RECOGNITION AND INSPECTIONS

## INTRODUCTION

This chapter is divided into three parts: The first, The Behavior-Based Safety Observation Process, contains some basic guidance for including a behavior-based observation process into a traditional hazard recognition and inspection effort. This relatively new approach focuses on unsafe acts and is an important supplement to traditional hazard recognition programs, which tend to focus on unsafe conditions. The second, Hazard Inspection Training, contains practical hazard recognition and inspection concepts and checklists to be used in providing hazard recognition training to general employees, to supervisors and managers, and to safety and health professionals. The third part of the chapter discusses OSHA and inspections conducted by OSHA compliance officers.

Like other information in this book, it will be necessary to tailor the contents of this chapter to the specific needs of your organization, but this chapter contains a wealth of information to aid in training general employees, supervisors, managers, and safety and health professionals in hazard recognition, control, and inspections.

## THE BEHAVIOR-BASED SAFETY OBSERVATION PROCESS

For decades the safety community has recognized that accidents are caused by unsafe acts and/or unsafe conditions. Traditional safety programs, however, have tended to focus on unsafe conditions, primarily because it is easier to define and identify conditions and because our codes, standards, and regulations tend to also focus on conditions.

## Background

Even though several approaches to observing and evaluating behavior (i.e., observable acts) have been made with varying degrees of success, the most effective efforts (in the author's opinion) have been based on the work done by Dr. Thomas Krause and Dr. John Hidley, founders of Behavior Sciences Technology (BST) of Ojai, California. In 1990, Krause and Hidley authored *The Behavior-Based Safety Process.* This book and the consulting and seminars offered by BST and others, have aided in providing a systematic, workable approach to including unsafe acts in our hazard recognition efforts.

In 1993, at the direction of Bruce Church, Assistant Manager for Environment, Safety and Health and Larry Skousen, Director of the Safety Division for the Nevada Operations Office of the Department of Energy (DOE), a steering committee was formed and four consultants were selected to develop a behavior-based process for use by the DOE Nevada Operations Office and its contractors.

The mission of the consultants was to develop a behavior-based observation process suitable for DOE operations at the Nevada Test Site and, hopefully, exportable to other DOE sites and operations. The process was to incorporate the best features of existing programs, improve upon existing efforts if possible, and attempt to achieve cost-effective methods of training and implementation.

The four consultants were Cliff McCullough of EG&G/Energy Measurements, Inc., Roger Herrod of Wackenhut Services Incorporated, Elaine Jacobson (later replaced by Bruce Williams) of Raytheon Services Nevada, and Joe Stephenson (this author), then of Reynolds Electrical and Engineering Company (REECo), Inc.

The information in this section is based upon the efforts of these consultants.

## What?

So what is the Behavior-Based Safety Observation Process? The Behavior-Based Safety Observation Process is a worker-managed process to improve safety in the workplace by observing and evaluating workplace behavior and providing feedback on that behavior (Fig. 8.1).

The process includes developing and implementing action plans to reduce or eliminate workplace acts that include "at risk" behavior and thus could result in accidents.

BEHAVIOR-BASED
SAFETY OBSERVATIONS

- WHAT - An observation and feedback process
- WHEN - During normal work activities
- WHERE - In the workplace
- WHY - To improve workplace safety
- HOW - By focusing on acts
- BY WHOM - Workers

**FIGURE 8.1** Fundamentals of behavior-based safety observations.

The characteristics of the Behavior-Based Safety Observation Process that set it apart from traditional safety efforts are as follows:

- The Behavior-Based Safety Observation Process focuses on observable *acts* whereas traditional safety efforts tend to focus on conditions.
- The Behavior-Based Safety Observation Process is a worker-managed process. It is imperative that the process be planned, implemented, evaluated, and upgraded by working level personnel.
- The Behavior-Based Safety Observation Process is positive and proactive. There is no disciplinary action for "at risk" behavior, only peer feedback and suggestions for improvement. Most of the feedback provided is positive feedback and reinforcement for performing work safely and properly.

### When?

Observations typically last 15 to 30 minutes and are generally performed during normal duty hours and during normal operations.

The process for determining the exact times that observations are made and the exact frequency is determined by individual observers based on guidance and direction from their implementation teams.

The Behavior-Based Safety Observation Process is ongoing. Once the process has been implemented, it becomes part of the normal way of doing business.

### Where?

Observations are made at the normal work location(s) of the person(s) being observed. This may be a plant location, an office, along a route, in multiple locations, or even "on the move."

The feedback session is normally given at the same location as the observation, but this can vary. For example, if the observation is in a noisy location or a harsh environment or "on the move," it may be desirable to conduct the feedback session in an office or break area if one is readily available. The location of the feedback session should provide privacy, and, to the extent practical, a degree of comfort.

The exact location to submit and/or input checklist data must be determined by each implementing team. This location should be convenient to the observers.

Generally the process is conducted within the same geographic boundaries as other work done by the organization.

### Why?

The purpose of the Behavior-Based Safety Observation Process is to improve workplace safety.

Peer observations and feedback provide awareness and incentive and aid in training workers in the safe and proper way to work (Fig. 8.2).

Traditionally, we tend to focus on the highly visible fatalities and other serious accidents. Yet we know that for every fatality there are many lesser injuries or accidents and many more unsafe acts that have the potential for injury.

We often attempt to rid ourselves of the accident iceberg by chipping away at the top. The only truly effective way to eliminate or even significantly shrink our iceberg is by warming the waters around it by creating a culture in which workers really do watch out for each other and by creating conditions that allow and promote open communications and feedback in a nonthreatening work environment in which an attitude of true concern for the well-being of fellow workers prevails.

### How?

The Behavior-Based Safety Observation Process improves safety by involving implementation teams comprised of workers to identify and prioritize work tasks and then to define what constitutes "safe" and "at-risk" behavior for each task. The implementation teams then develop observation checklists to aid in observing and evaluating behavior and recording the results.

Next, using the checklists as a guide, workers trained as observers observe behavior in the workplace and provide feedback on the observation. (*Note:*

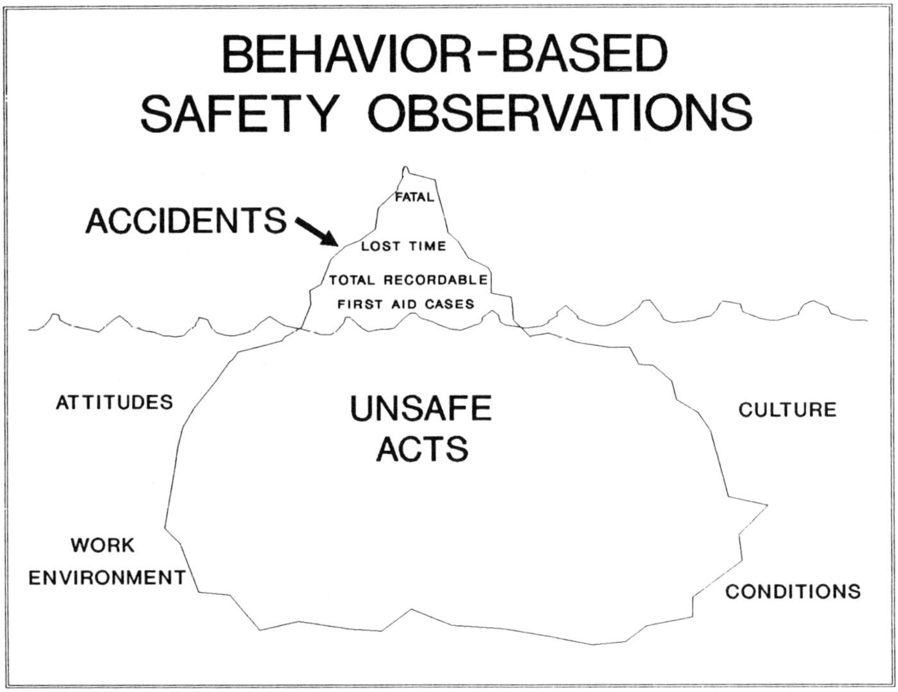


**FIGURE 8.2** Accident iceberg.

Initial training for implementation teams and observers is generally done by consultants or through public training workshops. Eventually, implementation teams and selected experienced observers may take over the task of training new observers.)

Immediate feedback is provided at the conclusion of each observation during a one-on-one discussion between the observer and the person being observed. Feedback is also provided via charts and reports provided, which show cumulative observations results and trends. Finally, cumulative data collected during observations are analyzed and action plans are developed to address serious or recurring problems.

These action plans are implemented and the results are evaluated. It may be necessary to modify checklists and/or observation schedules to ensure timely evaluation of action plan results.

Again, this is an ongoing process with active participants being systematically rotated so that all workers have a chance to participate.

Figure 8.3 summarizes the tasks to be performed by the workers who comprise the implementation team and/or the observers. (*Note:* During process startup, the implementation team will frequently also serve as the first group of observers.)

Checklists should not be too long. Generally a checklist should contain

**BEHAVIOR-BASED SAFETY OBSERVATIONS**

- THE PROCESS
  - Develop Observation Checklists
  - Train Observers
  - Conduct Observations
  - Provide On-the-spot Feedback
  - Compile Data
  - Print Charts
  - Develop Action Plans
  - Implement
- REPEAT/CONTINUE

**FIGURE 8.3** Behavior-based safety observation process.

only 10 to 20 items. It is very difficult for an observer to "watch for" more than a dozen or so specific behaviors during an observation.

Also, most checklists will have two parts: a list of general or generic behaviors common to many jobs and a list of job-specific behaviors that are unique to the specific work being done.

Table 8.1 is a generic checklist that was compiled by Cliff McCullough of EG&G/Energy Measurements. This list can be tailored to identify those behaviors most critical to the work being done.

Training to initiate the process requires consultants or public workshops to provide the implementation team members with a good understanding of the process, including the fundamentals, checklist development, observation and feedback training, data input, retrieval and analysis, and action plan development and implementation. The initial groups of observers may also require training or training assistance by consultants or public courses even though a goal of most efforts is to develop observer trainers within the implementation teams.

Generally, the fundamentals and checklist development workshop for the implementation teams should require 2 to 5 days, observer training should require 1 to 3 days, and data analysis and action plan development training for the implementation teams should take 3 to 5 days. The data analysis and action plan development training should take place 3 to 6 months after observations are initiated.

To get the process initiated or at least in the early planning stages, management orientations must be presented to introduce and sell the process to upper and mid-management. Even though it is essential that management play only a supporting role in developing and implementing the program, management support in terms of encouragement and resources is absolutely imperative.

**TABLE 8.1 Behavior-Based Safety Observation Checklist**

Observer: ________________

| Person 1 | Person 2 |
|---|---|
| Date: ________ | Date: ________ |
| Time: ________ | Time: ________ |
| Department: ________ | Department: ________ |
| Building: ________ | Building: ________ |
| Floor: ______ Area: ______ | Floor: ______ Area: ______ |

| 1.0 | Body Position | Safe | At-Risk |
|---|---|---|---|
| 1.1 | Eyes on Path | | |
| 1.2 | Eyes on Work | | |
| 1.3 | Stable Surface | | |
| 1.4 | Lifting | | |
| 1.5 | Overexertion | | |
| 1.6 | Line of Fire | | |
| 1.7 | Rushing/Shortcuts | | |
| 1.8 | Ascending/Descending | | |
| 1.9 | Pinch Points | | |
| 1.10 | Overextending | | |
| 1.11 | Path of Travel | | |
| 1.12 | Awkward/Cramped | | |
| 1.13 | Others | | |

**Definitions: 1.0 Body Position**

**1.1 Eyes on Path**
Is employee looking in the direction of travel?

Examples

a. If backing is necessary, visually check path or have someone direct traffic.
b. Don't be distracted by eating, reading, or looking away from the path being traveled.
c. Maintain eyes on path even when talking.
d. Watch for debris in path.
e. Watch for elevation changes.
f. Watch for overhead hazards such as pipes, scaffolding, or cranes.

**1.2 Eyes on Work**
Are the employee's eyes focused on the task being performed?

Examples

a. Watch for pinch points such as setting objects down with fingers beneath the object, blades or clamps shears, dyes of breaks or punches, or rotating equipment.
b. Watch for hot spots on materials after welding or cutting.
c. Watch for sharp chips when clearing machines.

**1.3 Stable Surface**
Is the work surface stable and secure?

Examples

a. Extension ladder is tied off at the top or another person is steadying at the base.
b. Ladder extends three feet above the area the employee is dismounting from.
c. Step ladder has spreaders locked and the base is on secure footing.
d. Scaffolding is plumb and base is on secure footing.

**1.4 Lifting**

Is correct lifting technique being used?

Examples

a. Moving heavy objects use work-saving devices such as dollies, forklifts, or another person.
b. Examine object to decide where and how to hold it.
c. Know where you will let the object down.
d. Stand close, squat down keeping back straight, grasp firmly, lift with legs, hold object firmly to body.
e. Turn with feet—do not twist back.

**1.5 Overexertion**

Worker does not extend physical capabilities.

Examples

a. Pushing or pulling file cabinets, etc. that are too heavy for one person to move or lift.
b. Working too long when full personal protective equipment is required.
c. Working too long in hot areas.
d. Working excessive periods without sufficient breaks.

**1.6 Line of Fire**

Is worker positioning body out of the direct path of materials and objects that may splash, fall, fly back, or act like a projectile.

Examples

a. Avoid working under suspended loads or where workers are working above.
b. Do not work in the path of flying debris (circular saw).
c. Make sure that pressure is relieved from lines when performing maintenance (air lines, steam, hydraulics, etc.).

**1.7 Rushing/Shortcuts**

Is worker taking shortcuts?

Examples

a. Running instead of walking.
b. Standing on stools, buckets, or chairs to reach upper shelves.

(*continued*)

**1.8 Ascending/Descending**

Is employee using correct techniques while ascending or descending stairs, scaffolding, or ladders?

Examples

a. Handrails are used while traveling up and down stairs.
b. Scaffold ladder used when climbing up and down scaffolding.
c. Employee is facing ladder when climbing up or down.
d. Three point contact is maintained when climbing up and down ladders.
e. Dismount carefully—do not jump.

**1.9 Pinch Points**

Is employee watching for placement of hands and the possible hazard to them from unexpected mechanical movement by stored energy or accidental start-up?

Examples

a. Meshing gears, pulley, or sheaves, or other moving parts.
b. Assuring that load will not shift when moving the load.
c. Using hand tools like hammers and wrenches.
d. Does employee make sure that body and extremities are clear of closing doors.

**1.10 Overextending**

Is employee reaching beyond personal ability?

Examples

a. Employee reaches beyond ability to turn valves.
b. Employee reaches excessively to pull objects or to maintain balance on ladders or scaffolds.

**1.11 Path of Travel**

Does employee look at the path of travel to ensure it is clear before moving into the path?

Examples

a. Before turning around in tight quarters does employee turn and look for obstructions first?
b. Does employee clear debris or spills from path of travel?
c. Employee walks in designated walkways.

**1.12 Awkward/Cramped**

Is employee using correct posture to perform task?

Example

a. Don't bend or stoop for extended periods without taking stretch breaks.

**TABLE 8.1** *(Continued)*

**1.13 Others**

| 2.0 | Work Environment | Safe | At-Risk |
|---|---|---|---|
| 2.1 | Walking/Working Surfaces | | |
| 2.2 | Barricades/Warnings | | |
| 2.3 | Obstructions | | |
| 2.4 | Housekeeping | | |
| 2.5 | Awkward/Cramped | | |
| 2.6 | Lighting | | |
| 2.7 | Ventilation | | |
| 2.8 | Others | | |

**Definitions: 2.0 Work Environment**

**2.1 Walking/Working Surfaces**
Use correct paths. Do not walk over objects or hazards in the path.

Examples
a. Oil, water, ice, pipes, other debris.
b. Take deliberate steps to know your footage.

**2.2 Barricades/Warnings**
Does employee use correct barricades or warnings for the work being performed?

Examples
a. Employee posts signs or guards for slippery surfaces, open pits or manholes, or laser/radiation areas.
b. Employee orally warns others when approaching work areas with suspended loads.
c. Employee sounds a warning when approaching work areas with operating vehicles.
d. Employee places welding screens up to protect visitors to the area.

**2.3 Obstructions**
Cautions are taken to minimize contact with obstructions in the work areas.

Examples
a. Sharp obstructions are padded before work continues.

**2.4 Housekeeping**
Is immediate work area and exit route free from potential hazards?

Examples
a. Floor kept free of oil, water, etc.
b. Is unused hardware picked up when job is complete?
c. Does employee leave the area as clean or cleaner than it was found?

*(continued)*

TABLE 8.1 *(Continued)*

**2.5 Awkward/Cramped**

**2.6 Lighting**
Is there sufficient light available to the work area?

Example
a Flash lights are used in poorly lit areas.
b. Employee provides additional lighting to work areas poorly lit.

**2.7 Ventilation**
Does employee ensure proper ventilation of work area?

Example
a. When welding, brazing, painting, etc. does employee use local exhaust or provide additional air moving equipment?

**2.8 Others**

| 3.0 | Tools and Equipment | Safe | At-Risk |
|---|---|---|---|
| 3.1 | Selection | | |
| 3.2 | Use | | |
| 3.3 | Condition | | |
| 3.4 | Vehicle Operation | | |
| 3.5 | Others | | |

**Definitions: 3.0 Tools and Equipment**

**3.1 Selection**
Is employee using the proper tool or equipment for the job being performed.

Examples
a. Use ladder sized correctly for the work.
b. Rigging tackle and forklifts sized correctly for lifts.
c. Insulated tools used when working on energized equipment.

**3.2 Use**
Is the employee using the tool or equipment correctly?

Examples
a. Cheater bars or pipes not used on wrenches.
b. Screwdrivers not used as putty knives, chisels, or pry bars or vice versa.

**3.3 Condition**
Is the tool or equipment in good working condition and clean prior to use?

Examples
a. Rigging tackle and handling equipment inspected before use.
b. Tools are clean and free of oil or grease to prevent dropping or slipping.

**TABLE 8.1** ***(Continued)***

**3.4 Vehicle Operation**

Is employee operating the vehicle in safe manner?

Examples

a. Does employee park the vehicle on flat surface to prevent doors from closing inadvertently?
b. Vision is not impaired by window fogging or icing.
c. Seat belts are worn by all occupants.
d. Speed limits are observed.
e. Lights, wipers, and brakes are in good working order.

**3.5 Others**

| 4.0 Personal Protective Equipment | Safe | At-Risk |
|---|---|---|
| 4.1 Eye and Face | | |
| 4.2 Head | | |
| 4.3 Hands | | |
| 4.4 Fall Protection | | |
| 4.5 Body Protection | | |
| 4.6 Shoes | | |
| 4.7 Hearing Protection | | |
| 4.8 Respirator Protection | | |
| 4.9 Others | | |

**Definitions: 4.0 Personal Protective Equipment**

**4.1 Eye and Face**

Is employee properly wearing appropriate eye/face protection that is in good condition and matches the hazards of the task?

Examples

a. Face shields and safety glasses when grinding, mono goggles in chemical areas, welding hoods with appropriate filters when welding.

**4.2 Head**

Is employee properly wearing an appropriate hard hat that is in good condition and matches the hazards for work being performed?

Examples

a. Area where overhead work is performed.
b. Area where danger of bumping into obstructions or protruding objects.
c. Fully insulated hard hat when exposed to electrical shock hazards.

**4.3 Hands**

Is employee properly wearing appropriate gloves in good condition and that match the hazards of the task?

*(continued)*

Examples
a. Rubber gloves in chemical areas.
b. Heavy leather or kevlar gloves when handling metal chips.
c. Leather gloves when handling rough wood, metal, wire rope, or arc welding.

**4.4 Fall Protection**

Is proper fall protection/prevention being worn properly, free of defects, and is the lanyard attached to a structure member that will support the employee in case of fall?

Examples
a. When working on open platforms above 4 feet high.
b. When working in buckets or aerial lifts.
c. Lanyard shorter than potential fall.
d. Buckles and straps securely fit to body.

**4.5 Body Protection**

Is employee wearing appropriate body protection to meet the hazards of the task?

Examples
a. No flammable clothing exposed to flame or sparks from arc welding or brazing.
b. Wear acid suit when working with hazardous chemicals.
c. Anti-C properly applied and sealed for potentially contaminated areas.

**4.6 Shoes**

Is employee wearing proper foot protection for the hazards of the task?

Examples
a. Hard toed shoes for work where objects dropping would cause injury to feet.
b. Insulated soles for work with electricity.
c. Oil-resistant soles in applicable environments.
d. Rubber boots in wet or chemical areas.

**4.7 Hearing Protection**

Is employee wearing appropriate hearing protection for the hazards of the task?

Examples
a. Wearing earplugs and/or earmuffs in high noise areas.
b. Hearing protection is being worn in areas that are posted or needed where noise level creates a danger to the worker's hearing?
c. If using power tools (saws, impact wrenches, grinders, compressors, generators, weed eaters, etc.).

**4.8 Respirator Protection**

TABLE 8.1 *(Continued)*

**4.9 Others**

| 5.0 Procedures | Safe | At-Risk |
|---|---|---|
| 5.1 Prejob Inspection/Planning | | |
| 5.2 Lack of Manpower | | |
| 5.3 Communication | | |
| 5.4 Complying with Lockout/Tagout | | |
| 5.5 Complying with Permits | | |
| 5.6 Others | | |

**Definitions: 5.0 Tools and Equipment**

**5.1 Prejob Inspection/Planning**

Is employee's equipment what is needed for the job and in working order?

Example

a. Has the work area been inspected for hazards such as condition of walk surfaces for holes needing barriers, sharp objects protected, rotating/moving equipment in adjacent areas, insect nests, asbestos or lead, lighting, adequate manpower, etc.

**5.2 Lack of Manpower**

Is adequate manpower present for the task being performed?

Examples

a. Carrying long ladders or lengths of pipe or stock.
b. Heavy or bulky loads being handled without sufficient assistance.
c. Moving heavy furniture or equipment.

**5.3 Communication**

Is verbal communication utilized effectively while work is being performed?

Examples

a. Fellow workers and others in the affected work area are warned of changes in the workspace configuration.
b. Employees communicate conditions of work progress as the work is being performed.
c. Employee requests assistance when work exceeds limitations such as quantity or weight.
d. Employees have adequate written or verbal instruction to safely perform the work at hand.
e. During flagging operations can the flagman be clearly seen.
f. Does the confined space monitor have communication ability to those in the space when out of sight?

*(continued)*

TABLE 8.1 *(Continued)*

**5.4 Complying with Lockout/Tagout**
Are proper lockout/tagout practices being performed.

Examples
a. Employee inspected all energy sources such as valves, pipes drained, electrical disconnect open, blocking in place?
b. Employee used personally assigned lock?
c. Employee tried system to ensure lockout/tagout?

**5.5 Complying with Permits**
Does the work being performed require a permit and is it being followed?

Examples
a. Employee ensured the permit is approved.
b. There is appropriate safety equipment on the job (fire extinguisher)?

**5.6 Others**

| 6.0 Apparel/Accessories | Safe | At-Risk |
|---|---|---|
| 6.1 Hair | | |
| 6.2 Clothes | | |
| 6.3 Jewelry | | |
| 6.4 Others | | |

**Definitions: 6.0 Apparel/Accessories**

**6.1 Hair**
Is hair tied back or under netting to prevent being caught in rotating equipment?

**6.2 Clothes**
Is clothing fitted properly and not hanging loose to get caught in rotating equipment?

Examples
a. Sleeve cuffs are closed and incapable of being caught on moving equipment?
b. Coveralls are not frayed excessively when worn around open flames or welding activities?

**6.3 Jewelry**
Is jewelry worn in the work area that could become entangled in equipment or conduct electricity?

Examples
a. Employee is not wearing rings or metal band watches or bracelets while working on electrical circuits or batteries?
b. Employee is not wearing loose necklace that could get caught in equipment?

**TABLE 8.1** ***(Continued)***

**6.4 Others**

| 7.0 | Housekeeping | Safe | At-Risk |
|---|---|---|---|
| 7.1 | Floor | | |
| 7.2 | Equipment | | |
| 7.3 | Storage of Materials | | |
| 7.4 | Others | | |

**Definitions: 7.0 Housekeeping**

**7.1 Floor**
Does employee clear floor of debris or spills that would cause a trip hazard or fire hazard?

**7.2 Equipment**
Does employee properly store equipment being used such that it is not a trip or bump hazard or a falling hazard? Is equipment cleaned to prevent slipping or gripping hazard from oily or wet surfaces and to prevent a fire hazard?

**7.3 Storage of Materials**
Does employee properly store paints and flammable materials when finished being used? If equipment must be stored in walkways temporarily, does employee place warnings or barriers around them?

**7.4 Others**

| 8.0 | Office Safety | Safe | At-Risk |
|---|---|---|---|
| 8.1 | Housekeeping/Storage | | |
| 8.2 | Ergonomics | | |

**Definitions: 8.0 Office Safety**

**8.1 Housekeeping and Storage**
Is material stored on top of cabinets or shelving where it presents a falling hazard?

**8.2 Ergonomics**
Does employee take stretch breaks while working extended periods at word processing desk tasks or repetitive motion tasks?

In addition, orientations and "buy in" meetings should be held for all members of the organizations participating in the process. These may be presented by consultants, but frankly are most effective when given by workers from the implementations teams with assistance in preparation but minimal participation by consultants. Managers should be present to express support for the effort but should also play a minimal role.

## By Whom?

The Behavior-Based Safety Observation Process is run by working level employees for working level employees. Workers within specified work units (crews, branches, sections, etc.) develop the observer checklists, train observers, conduct the observations, provide feedback, maintain the data, analyze the data, develop action plans, and implement action plans.

Other specific responsibilities are as follows:

### Managers/Supervisors

- Provide resources and support

### Safety Community

- Maintain current programs
- Ensure compliance and safe conditions

### Training Community

- Provide Train-the-Trainer instruction as needed
- Maintain individual training records

### Steering Committees

- Provide broad guidance and support
- Ensure appropriate uniformity and consistency

### Consultants

- Serve as project managers during planning and development phases
- Conduct initial training
- Serve as consultants during implementation

### Implementation Teams

- Develop definitions and observation checklists
- Select and train observers
- Maintain data and generate reports
- Coach observers
- Analyze data and develop action plans
- Implement action plans

### Observers

- Conduct observations and provide on-site feedback

- Document observations and provide data input
- Assist implementation teams

Even though the goal is for the effort to become self-supporting and independent as rapidly as practical, external consultants and/or outside training are almost always required for successful implementation of a behavior-based safety observation process.

Even though Behavior-Based Safety Observations are not required by OSHA or any other code, standard, or regulation, they are highly recommended.

This process is an excellent way to demonstrate leadership in safety management and to transfer ownership of safety responsibility to the line organization.

Personal experience and that reported by others indicates that this process can produce very significant positive results.

Figure 8.4 contains a summary of Behavior-Based Safety Observation attributes.

For the latest information on assistance available for implementing the Behavior-Based Safety Observation Process, contact:

Dr. Joe Stephenson, CSP
West Coast Regional Manager
COBRO Corporation
8429 Honeywood Circle
Las Vegas, Nevada 89128

Phone: (702) 363-5760
Fax: (702) 256-5380

**BEHAVIOR-BASED**
**SAFETY OBSERVATIONS**

- SUMMARY:
  - Process focuses on acts, not conditions
  - Process is run by and for workers
  - Very proactive in that it addresses all unsafe acts, not just those which result in accidents
  - Process provides positive feedback and reinforcement since most acts are safe acts
  - There is no discipline or negative consequences associated with the process
  - Process is an excellent employee awareness and education tool, based on Total Quality Management principles of empowerment, involvement, measurement, and continuous improvement

**FIGURE 8.4** Summary of attributes.

## HAZARD INSPECTION TRAINING

### Continuing and Scheduled Safety Observation Tours

The Occupational Safety and Health Act of 1970 makes it necessary for every plant to establish some type of self-inspection and hazards correction program.

Organizing and implementing an OSHA compliance program can be a monumental task if not approached systematically. If undertaken with the philosophy that standards compliance can be tailored to a plant, as well as vice versa, instituting a compliance program need not be overwhelming. Compiling standards abstracts will make the job easier.

The first step is to collect and organize needed information, including the federal (OSHA) standards, state and local codes, the National Electrical Code, and applicable fire protection codes. Some type of easy-to-use photographic equipment will also aid the gathering and documenting of data on plant conditions.

When these materials have been gathered, safety inspection checklists (Table 8-2) and safety and health self-inspection reports (Table 8-3) can be started. Here are some tips that can be helpful.

Keep the job as simple as possible and do not try to do it all at once. OSHA publication 2201, "General Industry: OSHA Safety and Health Standards Digest," can be invaluable when organizing the program. It is available from the nearest OSHA office.

Plan on more once-over-lightly inspections with later in-depth follow-ups. See Table 8.4 for a list of possible problems to be inspected.

### High Accident Areas and Operations Emphasized

Proceed on a "worst first" basis. There are two reasons for doing this: any serious or fatal accident must be immediately reported to OSHA and may trigger an official inspection, and the likelihood of employee complaints will be reduced.

In terms of cost effectiveness, far more accidents will be prevented (therefore, dollars saved) by concentrating attention on the worst accidents first. Analysis of the company's accident history should reveal where and how and when its most costly accidents are occurring. This could be an occasional accident that is costly in terms of death, disability, and/or many lost person days, or it could be an inexpensive accident that rarely results in more than one or two lost person days but occurs frequently.

Whether it is a hazardous area where accidents are taking place or a hazardous operation that is the site of costly accidents, the company will realize greater return on its investment in safety training by concentrating its attention on the "worst first."

**TABLE 8.2 Safety Inspection Checklist**

Name of Facility ______________________ Date of Survey __________
Address (Street)________________________________________________

________________________________________________________________
City State Zip Area Code and Telephone

| | Yes | Partly | No |
|---|---|---|---|
| *A. Management* | | | |
| 1. Has set the safety policy for the facility | | | |
| 2. Exhibits an active interest in the conduct of an effective safety program | | | |
| 3. Budgets funds for long- and short-range safety program items and projects | | | |
| 4. Holds the superintendent and staff responsible for maintaining a safe environment and for providing a meaningful safety program | | | |
| *B. The Superintendent* | | | |
| 1. Exhibits leadership in the facility safety program | | | |
| 2. Holds the staff responsible for an effective safety program | | | |
| 3. Plays an active role in the operation of the executive safety committee | | | |
| 4. Reports to management on the safety climate and problems | | | |
| 5. Requires an accident reporting and records system | | | |
| 6. Has organized an executive safety program | | | |
| *C. The Plant Manager* | | | |
| 1. Plans and directs an effective safety program | | | |
| 2. Conducts meetings of the executive safety committee, at least quarterly | | | |
| 3. Provides the superintendent with a copy of minutes of all safety committee meetings | | | |
| 4. Requires that employees be furnished, without cost to themselves, all necessary personal protective safety equipment | | | |
| 5. Requires that all supplies, materials, and equipment reflect accepted safety requirements | | | |
| 6. Requires that the preventive maintenance program be integrated with appropriate safety requirements | | | |
| 7. Requires that foremen maintain safe working conditions and practices in the operations under their supervision | | | |
| 8. Provides for the safety training of all foremen | | | |
| 9. Requires periodic safety inspections of all work areas by foremen | | | |

(*continued*)

**TABLE 8.2** ***(Continued)***

| | Yes | Partly | No |
|---|---|---|---|
| *D. Building* | | | |
| 1. Structurally sound | | | |
| 2. Ramps provided | | | |
| 3. Entrances—one primary entrance for wheelchairs | | | |
| 4. Doors and doorways—adequate width | | | |
| 5. Doors and doorways—landing platforms meet standards | | | |
| 6. Doors and doorways—door sills meet standards | | | |
| 7. Stairs meet standards | | | |
| 8. Steps have rounded nosings | | | |
| 9. Stair handrails are standard | | | |
| 10. Stair handrails extend beyond top and bottom steps 18 inches | | | |
| 11. Stair step design is standard | | | |
| 12. Floors have nonslip surface | | | |
| 13. Floors are free of holes and projections | | | |
| 14. Floors on a given story have common level | | | |
| 15. Toilet rooms are designed for use by handicapped persons | | | |
| 16. Water fountains are designed for use by handicapped persons | | | |
| 17. Public telephone is installed for use by handicapped persons | | | |
| 18. Elevators are accessible for use of handicapped persons | | | |
| 19. Switches and controls for light, heat, fire alarms, and all similar controls of frequent or essential use are placed within reach of person in wheelchair | | | |
| 20. Facilities in building that would be hazardous for the blind should be specifically identified for recognition by blind persons | | | |
| 21. Visual and audible warning signals are interrelated for benefit of the blind and deaf | | | |
| 22. Exit signs meet standards | | | |
| 23. Aisles are defined with marking lines on floor | | | |
| 24. Aisles are wide enough to accommodate two wheelchairs | | | |
| 25. Protruding hazards marked in accordance with safety color code | | | |
| *E. Grounds* | | | |
| 1. Safe conditions for entering and existing privately owned vehicles | | | |
| 2. Hazards on facility grounds have been eliminated or controlled | | | |
| 3. Public access roads for employees using public transportation | | | |
| 4. Parking lots used by employees meet standards | | | |

**TABLE 8.2** ***(Continued)***

| | Yes | Partly | No |
|---|---|---|---|
| *F. Storage* | | | |
| 1. Storage areas and rooms are orderly | | | |
| 2. Stored material are handled safely | | | |
| 3. Mechanical materials handling equipment is safely operated | | | |
| *G. Machinery* | | | |
| 1. Machinery moving parts are provided with protective guarding | | | |
| 2. Instruction is given on safe use of power tools | | | |
| 3. Supervision follow-up on power tool safety instruction is adequate | | | |
| 4. Point of operation guards have been installed | | | |
| 5. Point of operation guards are used by personnel | | | |
| *H. Lighting* | | | |
| 1. Illumination levels for seeing tasks meet standards | | | |
| 2. Illumination sources in hazardous areas meet standards | | | |
| 3. Illumination on ramps and stairways is satisfactory | | | |
| *I. Ventilation* | | | |
| 1. General ventilation of work areas is satisfactory | | | |
| 2. Hazardous fumes are removed to safe discharge points | | | |
| *J. Personal Protective Safety Equipment* | | | |
| 1. Proper types of equipment are provided by management | | | |
| 2. Supervision follow-up ensures use by personnel | | | |
| *K. Compressed Gas Cylinders* | | | |
| 1. Storage and handling are in accordance with safety standards | | | |
| *L. Electrical Wiring System (Permanent and Temporary)*[a] | | | |
| 1. Wiring meets needs of the facility | | | |
| 2. Sufficient number of convenience outlets | | | |
| 3. Overloading of circuits is controlled | | | |
| 4. Use of extension cords is controlled | | | |
| 5. Emergency switches are spaced and functioning satisfactorily | | | |
| *M. Elevators* | | | |
| 1. Elevators are inspected and maintained on a periodic and scheduled basis | | | |
| 2. Elevator cars and shaftway openings are protected by doors or gates interlocked with elevator controls. | | | |

(*continued*)

**TABLE 8.2** ***(Continued)***

| | Yes | Partly | No |
|---|---|---|---|
| *N. Pressure Vessels* | | | |
| 1. Boilers are inspected and approved by competent inspectors at least annually | | | |
| 2. Air receivers are inspected and approved by competent inspectors at least annually and condensate is drained regularly. | | | |
| *O. Electrical Equipment* | | | |
| 1. Equipment is correctly fused | | | |
| 2. Equipment is properly maintained and cleaned | | | |
| 3. Portable electrical equipment is grounded | | | |
| *P. Fire Hazards* | | | |
| 1. Housekeeping is generally acceptable | | | |
| 2. Housekeeping in above-normal risk areas is adequate | | | |
| 3. Waste containers are adequate in design and availability | | | |
| 4. Waste containers are properly disposed of at end of each day | | | |
| 5. Waste paper and rags are disposed of regularly | | | |
| 6. "No Smoking" signs are posted and observed | | | |
| 7. Flammable liquids are safely dispensed | | | |
| 8. Flammable material is stored safely | | | |
| 9. Wiring in hazardous areas meets National Electrical Code | | | |
| 10. Spray painting booths meet standards | | | |
| 11. Spray painting areas are kept free of accumulations | | | |
| *Q. Fire Prevention Controls* | | | |
| 1. Fire extinguishers are properly serviced and maintained | | | |
| 2. Personnel periodically are given instruction in the use of extinguishers and fire hose | | | |
| 3. Fire instructions for each floor have been issued and discussed with staff and clients | | | |
| 4. Sprinkler heads are unobstructed | | | |
| 5. Fire doors are functional | | | |
| 6. Fire escape stairway is free of obstruction | | | |
| 7. Fire aisles are kept clear | | | |
| *R. Medical Program* | | | |
| 1. Doctor on duty and in charge of medical program | | | |
| 2. Registered nurse on duty | | | |
| 3. First aid room and facilities available on premises | | | |
| 4. Contract medical services provided (alternative to items 1 and 2) | | | |

**TABLE 8.2** ***(Continued)***

| | Yes | Partly | No |
|---|---|---|---|
| *S. Personal Protective Equipment* | | | |
| 1. Each person exposed to eye hazards is furnished with personal goggles or a face shield that meets standards | | | |
| 2. Each person exposed to respiratory hazards is furnished with a personal respirator that meets standards | | | |
| *T. Special Items* | | | |
| Note (Add special items if desired). | | | |

Name(s) and Title(s) of Person(s) Involved in Survey:

Must conform to the requirements of the National Electrical Code.

[a] Must conform to the requirements of the National Electrical Code.

## Planning a Formal Inspection for Safety and Health Hazards

### *Objectives of Planning a Formal Inspection*

- To recognize all undesired practices and conditions that could result in personal injury or the impaired health of the employees, or damage to equipment, and to identify relevant standards.

**TABLE 8.3 Safety and Health Self-Inspection Report**

Area Inspected ______________ Date ______________ Inspector ______________

| Code | Hazards | Corrective Action |
|---|---|---|

Analysis and Comments ______________________________

*Source:* "Principles and Practices of Occupational Safety and Health: A Programmed Instruction Course," OSHA 2215; Student Manual Booklet 3, U.S. Department of Labor, Washington, D.C., p. 45.

**TABLE 8.4 List of Possible Problems to Be Inspected**

| | |
|---|---|
| Acids | Hoists |
| Aisles | Hoses |
| Alarms | Hydrants |
| Atmosphere | Ladders |
| Automobiles | Lathes |
| Barrels | Lights |
| Bins | Mills |
| Blinker lights | Mists |
| Boilers | Motorized carts |
| Borers | Piping |
| Buggies | Pits |
| Buildings | Platforms |
| Cabinets | Power tools |
| Cables | Presses |
| Carboys | Racks |
| Catwalks | Railroad cars |
| Caustics | Ramps |
| Chemicals | Ramps |
| Claxons | Respirators |
| Closets | Roads |
| Connectors | Roofs |
| Containers | Safety devices |
| Controls | Safety glasses |
| Conveyors | Safety shoes |
| Cranes | Scaffolds |
| Crossing lights | Shafts |
| Cutters | Shapers |
| Docks | Shelves |
| Doors | Sirens |
| Dusts | Slings |
| Electric motors | Solvents |
| Elevators | Sprays |
| Explosives | Sprinkler systems |
| Extinguishers | Stairs |
| Flammables | Steam engines |
| Floors | Sumps |
| Fork lifts | Switches |
| Fumes | Tanks |
| Gas cylinders | Trucks |
| Gas engines | Vats |
| Gases | Walkways |
| Hand tools | Walls |
| Hard hats | Warning devices |

*Source:* "Principles and Practices of Occupational Safety and Health: A Programmed Instruction Course," OSHA 2213, Student Manual Booklet 1, U.S. Department of Labor, Washington, D.C., p. 40.

- To evaluate and record the general degree of hazard associated with each item as a guide to setting up remedial and preventive action priorities.

### *Three Principles in a Planned Safety and Health Inspection*

When performing a planned inspection, keep in mind the following principles:

- OSHA standards require inspections in certain situations.
- Normal wear and tear eventually brings trouble.
- People make mistakes.

The regularly scheduled inspections should come at planned times during the year, month, week, or day, depending on the particular circumstances, or on the item(s) to be inspected. For example, it is generally a good idea to check for cleared work areas (good housekeeping) daily, whereas fire extinguishers are usually checked monthly. Therefore it is necessary to decide on when, or at what times, certain items need to be inspected.

### *Poor Planning Leads to Poor Results*

- The planning of a formal inspection is very important.
- If planning is poor, results will also be poor.

### *Inspections Should Be Well Organized*

To be effective, any inspection program must be well organized and planned in a way that ensures complete and timely inspection of all facilities. This means having regular, systematic inspections, in addition to the ongoing ones that always should be in process (i.e., constantly looking for hazards during the daily routine).

Daily inspection alone, even though organized, planned, and embodying a checklist, will not identify all the hazards in a work area. Periodic, planned inspections will be most beneficial in identifying additional hazards, as well as helping to decide what steps must be taken to control them.

### *Questions to Answer Before Performing an Inspection*

To effectively plan for a formal inspection, management should always require that written answers to the following key questions be submitted prior to the inspection:

- What problem will be inspected?
- What are the critical factors for each problem to be inspected?
- What conditions will be inspected?

### *Two Types of Planned Safety and Health Inspection*

There are two types of planned inspection that a supervisor might use to good advantage:

- Inspection of the supervisor's own work area
- "Audit" inspection of another supervisor's work area

The first type of inspection is vitally important because the employer is responsible for what happens in that work area, and the supervisor's position makes him or her a direct representative of management. Also, the supervisor is probably the person most familiar with all aspects of the work area and its relationship to the elements of hazard analysis previously discussed.

### *Audit Inspections Help Everyone*

Audit inspections are planned in the same manner as inspection of one's own work area, which is described later. The only difference is that in an "audit" inspection, one supervisor inspects another supervisor's work area. An important point to keep in mind when performing an audit inspection is that the goal is to *help* the other supervisor by discovering problems he or she may have missed.

### *Benefits of an Audit Inspection*

Audit inspections produce a two-way benefit. First, they help two supervisors do a good turn for each other and for their employees. Second, they help each supervisor gain a new perspective. After looking over someone else's work area, a supervisor may come back to his or her own area with added or better ideas for improving that operation. As can be seen, there *are* benefits to be gained by performing audit inspections.

### *Objectivity Is Important*

The importance of the second type of inspection, "audit inspection of another work area," may not be as immediately obvious as the first type. Oddly enough, one of the greatest advantages a supervisor has in a facility inspection—familiarity with the employees, equipment, and machines in the area—can be a disadvantage. Personal involvement with a work area can rob a supervisor of one of the most critical ingredients of a good safety and health inspection, namely, objectivity. The less objective a supervisor is, the more likely it is that he or she will miss dangerous hazards.

Being human, people have trouble remaining totally objective if they are personally involved in a situation. Almost all people tend to lose some degree of objectivity in repeatedly inspecting their own work area. One factor is the tendency to overlook anything not included on the checklist; thus the checklist can become a disadvantage as well. Checklists are guides and should never be considered as final, all inclusive, or never changing.

### *Hazards Tend to Accumulate*

Many safety or health problems tend to accumulate very gradually. This gradual change is hard to recognize if the same person who does the inspection works day after day in the same area.

### *Formal Inspection*

For purposes of discussion, the audit inspection and the planned, periodic inspection are termed formal inspections. They occur at certain times and, as will be seen later, they give a great amount of detailed and useful information. Since formal inspections are usually the most thorough, they are very important. Remember, however, that these formal inspections cannot replace daily inspections by supervisors.

*Note:* One point to keep in mind is that each employee should fix the hazards that he or she has the authority to correct. Do not wait until someone is injured or becomes ill before exercising authority that has already been given.

## OSHA Standards Used as Inspection Guides

Begin to organize the checklists by referring to the first page of the standards—actually, a table of contents. Note that some of the subparts may not apply to all plants and can, therefore, be ignored in certain cases. Become familiar with, and evaluate, the individual subparts. For example, consider the following:

**Subpart A—General.** This section applies to all plants but contains no direct references to actual plant requirements. It is worth mentioning that some of the general standards widely publicized as "scare" items have no basis in fact. Rather, they have been fabricated by opponents of OSHA and the OSHAct. Sections 1910.3 and 1910.4 outline opportunities available if a standard can be documented as unworkable.

**Subpart B—Adoption and Extension of Established Federal Standards.** If any of these standards apply to your plant, obtain copies of the applicable standards.

**Subpart R—Special Industries.** Plants not within the industries listed are not obligated to comply with any of these standards. However, some of them might be considered useful for other industries establishing a very thorough compliance program.

Once the standards have been generally studied, they must be examined in detail. With a transparent yellow marker, check the standards that specifically apply and will require immediate action. (The yellow marks will not show on photo copies that are made later.) It is also a good idea to remove the staples from one set of the OSHA standards and keep it in a loose-leaf binder so that pages can be easily removed, copied and replaced.

Next, make a copy of each page on which standards have been marked for attention, cut out the needed paragraphs, and glue them on the left-hand side of a blank sheet of paper. This arrangement leaves room for note making. Cutting out paragraphs and remounting them permits editing of the unnecessary material and the bringing together of related standards that may not appear consecutively in the original publication. The finished OSHA standards abstract is now ready to be copied and distributed as necessary.

A copy of each standards abstract, including any instructional notes, should be retained in follow-up files. These files can be arranged by department involved or by standard. It may be advantageous to use both systems with cross-referencing.

Using an abstract system to organize OSHA compliance paperwork offers many advantages. An abstract does the following:

- Helps direct attention and establish priorities.
- Allows various related standards to be combined
- Legally justifies action to be taken by reproducing the exact wording of standards
- Educates personnel who receive it
- Provides basic documentation for the compliance program
- Reduces time needed to explain the why's and how's of a compliance requirement

### Simulated Plant Inspection

Now let us accompany a hypothetical OSHA compliance officer on a walk-around inspection of a workplace. This is an opportunity to review some potentially hazardous conditions that are found in many workplaces all over the United States. As the simulated inspection is conducted, we will attempt to create a mind's-eye picture of the situation, describe the hazard(s), and consider the various alternatives to abating each of them.

#### Situation 1

Perception: An employee, probably assigned to the maintenance department, is using a portable piece of cleaning equipment to work on the floor of a manufacturing area.

Hazard: Close-up inspection reveals that one of the mechanical guards is missing from the equipment. Additionally, our sound level meter indicates a noise problem.

Abatement: Locate or fabricate the missing guard and request that it be installed. Instruction regarding future actions to keep guards in place may be needed as well. An attempt should be made to muffle the noise sufficiently to bring it into compliance with legal limits. If this cannot be accomplished, it may be necessary to replace the equipment with a model that does not offer this noise hazard.

**Situation 2**

Perception: An employee is climbing a portable wooden ladder in a backroom storage area.

Hazard: There are no railings on the landing platform and the toe board is missing. The ladder is too short at the point of access to the landing (it should extend 36 inches). The employee is facing the ladder as he should, however he is observed to be supporting himself with only one hand. The other hand appears to be used for carrying tools or materials.

Abatement: The department foreman should be advised of the physical hazard (i.e., the missing toeboard), with a recommendation that this requirement, as well as the ladder and climbing procedures, be a topic for continued inclusion in training sessions.

**Situation 3**

Perception: In checking the floor areas, an open grease pit in the company garage is noted.

Hazard: Although sockets for guard rails exist, railings are not in use to inhibit personnel from falling into the pit. Additionally, we find an oily rag on the floor and generally poor housekeeping.

Abatement: The rails should be located and installed, and the area cleaned up. This appears to be a point for regular follow-up during safety meetings.

**Situation 4**

Perception: We begin to walk down a stairwell and note a 55-gallon drum at the base of the stairs.

Hazard: The stairway, therefore personnel passage, is partially blocked by the drum.

Abatement: The drum should be removed and personnel reminded to keep possible fire exits clear of obstructions *AT ALL TIMES.*

**Situation 5**

Perception: As we continue down the next flight of stairs, a large pedestal floor fan blocks our way at the base of the stairs.

Hazard: Same as Situation 4.

Abatement: Same as Situation 4.

**Situation 6**

Perception: As we walk by an area where carpenters are remodeling a suite of offices, we notice a large hole in the floor.

Hazard: The hole should be safely covered and/or surrounded by temporary barriers, with appropriate signs to reduce the likelihood of someone tripping or falling into or through the hole.

Abatement: This is another reminder item for inclusion in safety meetings.

**Situation 7**

Perception: We step out onto the top level of the building. The equipment and ventilators on the roof have to be maintained regularly, so there is frequent need for employees to be here. The guard rail at the perimeter is good, however, poor housekeeping (see Hazard) could cause its strength to be put to the test.

Hazard: Many items have been carelessly strewn about this level and offer a real tripping/falling hazard.

Abatement: The area should be cleaned up, and all serviceable items should be placed in the appropriate storage areas. Unserviceable materials should be carefully stored to await repair or appropriate disposition. This item should be a point for discussion at safety meetings.

**Situation 8**

Perception: As we enter a large production area with many machines (cranes, drills, saws, etc.), we recognize the presence of oily greasy floor.

Hazard: Oily or greasy floors constitute hazards that bring on falls, and falls cause injuries ranging from fractures to twisted backs or even worse. Additionally, fire potential is significantly increased with oily floors.

Abatement: The walking areas should be degreased or covered over immediately and personnel traffic removed until the hazard is eliminated. Operators and maintenance personnel should be made aware of their responsibilities in keeping this hazard to a minimum.

**Situation 9**

Perception: As we continue our inspection, we note a wheel-mounted fire extinguisher positioned in its floor-marked storage area.

Hazard: A 55-gallon drum has been carelessly left in front of the extinguisher storage area blocking access to it. In addition, the area is not clearly marked so that personnel can locate it rapidly in an emergency.

Abatement: The area should be specifically identified as containing a fire extinguisher. All personnel should be cautioned against blocking access to any emergency equipment.

**Situation 10**

Perception: Next we see a wall-mounted fire extinguisher posted on the inside of a large door jamb.

Hazard: Closer observation reveals that a bundle of welding rods lies on the floor below the extinguisher. A spectacular spill could result if someone ran up to grab the extinguisher and slipped on those rods. Of course, this would happen only if the extinguisher could be found, and there are no wall markings on either side of the door jamb to call attention to its presence.

Abatement: Maintenance personnel should be regularly reminded of their safety-related responsibilities, and in this situation they can be shown that they have not been discharging their responsibilities. This applies to both job cleanup and fire extinguisher placement and wall markings.

**Situation 11**

Perception: As we walk through one of the ground floor exits, we find a storage area for compressed gas cylinders. Some are setting on their base and secured by a chain, but most are either leaning against a wall or some less secure support or are flat on the ground.

Hazard: Compressed gas cylinders must be stored and used safely. All cylinders should be identified and segregated according to whether they are full or empty. Proper storage requirements must be implemented.

Abatement: Personnel with the responsibility for caring for this material should be advised of the consequences of neglecting storage requirements. Sufficient areas should be made available to allow for meeting these requirements.

**Situation 12**

Perception: Next we approach a specific metal cylinder used to store compressed gases.

Hazard: Close-up inspection reveals an unknown substance that seems to have come from inside and is surrounding the cylinder cap.

Abatement: Isolate the storage area and do not move the cylinder. Contact the supplier to identify the level of hazard and the recommended method of elimination.

**Situation 13**

Perception: As our walk-around inspection continues, we see a maintenance crew storage and work room that is immediately dubbed "the snakepit."

Hazard: What we have found is a room that seems to exemplify a careless attitude toward good housekeeping and safe practices on the part of the people who use this area. We see tools strewn on raised work surfaces as well as the floor. Additionally, we see hundreds of feet of heavy electrical cord and rope, some tied in loose bundles, haphazardly located throughout the area. Some power drills with bits inserted are plugged in and resting on the floor near the base of the raised working surfaces.

Abatement: It is not enough to settle for cleaning this area up. The appearance of this room is a clear indicator of a need for additional safety training as well as a better attitude about responsibilities for safety. This one will take a while.

**Situation 14**

Perception: Next we find an electric motor that drives a machine through a mechanical power transmitter.

Hazard: Here we have a chain drive transmission exposed without any form of guarding.

Abatement: Locate and reinstall the original guard. If it cannot be found, one should be fabricated to provide necessary protection from fast-moving pinch points.

**Situation 15**

Perception: A look outside the plant reveals a parked truck on an inclined driveway adjacent to the dock.

Hazard: If the truck begins to move while personnel are working in or near it, there is a high probability of injury or death.

Abatement: Vehicles parked in a dock area should be chocked. Drivers and materials handlers should be so advised, and a warning sign should be posted as a reminder if there is not one there already.

### Situation 16

Perception: In the production area we see a medium to high-speed production line machine that could cause its operator to lose his or her fingers or entire hand.

Hazard: We see that the guarding system does not sufficiently inhibit the machine's operator from getting the hands too close to the machine while it is operating.

Abatement: The machine requires immediate guarding. Until such guarding is provided, however, operators of the machine should not be expected to expose themselves to possible injury. As a part of their safety training, all personnel should receive frequent reminders to call situations like this to management's attention so that hazards can be expediously eliminated.

### Situation 17

Perception: One wall has been established as a storage point for hanging slings that are not currently in use but are always on call.

Hazard: At first glance, the situation looks acceptable. Unfortunately, a closer look reveals that most of the slings are kinked and frayed, thus exposing their users to accidents due to broken slings and dropped loads.

Abatement: A regular inspection and maintenance program should be established to eliminate the high risks associated with a kinked or frayed sling.

So far, we have seen some cases of standards related to walking and working surfaces, guarding, fire protection, housekeeping, and other areas.

Now let us introduce another ingredient—occupational health problems. This is an area that at last is receiving the long overdue recognition and attention it deserves as a vital component of the total safety and health system.

Many, if not most, work situations have actual or potential health hazards. Some of these are illustrated in the continuing series of situations.

**Situation 18**

Perception: Over in the plating area, where employees are regularly exposed to strong acids as a part of their daily operations, we see an eyewash fountain.

Hazard: As a precaution, we check the operation of the fountain by pulling the chain, which should cause the water to flow freely. Much to our dismay, we note that this emergency device fails to function. In the event of an emergency, an employee who had immediate need of the eyewash fountain would be out of luck.

Abatement: A regular maintenance and checkout program should be established to ensure the immediate availability of this equipment in the event of an emergency.

**Situation 19**

Perception: Over in the locker room, we find an industrial sink. Adequate washing facilities are, we recall, also important for controlling skin diseases. Dermatitis is a common type of occupational illness.

Hazard: Stepping on the operation ring near the floor to verify the utility of the sink, we see that only a trickle of water is coming out of the pipes.

Abatement: A regular maintenance and operational checkout program should be established to ensure that washing capability is always available.

**Situation 20**

Perception: Moving through the welding area, we note the presence of arc welding.

Hazard: Arc welding creates exposure to ultraviolet light, which may result in eye damage. It similarly involves exposure to air contaminants, since welding on mild steel generates iron oxide fumes. When the fumes pass in and around the welder's face shield, he or she may breathe in harmful amounts.

Abatement: A local exhaust system should be installed to effectively remove fumes resulting from arc welding. If the welder is properly positioned (i.e., out of line of the main fume draft), exposure is greatly reduced and general workplace ventilation may be sufficient to adequately control exposure.

**Situation 21**

Perception: Even before we enter a punch press area and see the signs identifying it as a high noise area, we can hear the tremendous sounds of the impact of metal on metal.

Hazard: At the outset, we are not aware of what effort has already gone into attempting to eliminate or reduce the noise problem. However, several employees are working in the area, and all but one are wearing personal ear protection.

Abatement: First, a scientifically controlled analysis of the noise intensity in the area should be conducted. Assuming there is a problem, an attempt should be made to physically eliminate or reduce it. Effective engineering control is considered to be the only real solution to excessive noise problems. The use of personal ear protection should be considered to be a temporary expedient for use only when engineering or administrative controls cannot solve the problem. Furthermore, it appears at first glance that not all the employees consider the excessive noise level to be a problem. This is a topic that should receive regular attention as a part of the safety training program.

Additionally, the regular measurement of the noise exposure in an operation where the noise may be hazardous to hearing should be incorporated into the overall safety program. The use of either ear muffs or ear plugs for personal protection should be an individual's own choice. Together they may reduce noise exposure as much as 30 dBA (decibels on the "A" scale of a noise measuring device) Accomplishing this degree of attenuation, however, depends, to a great extent, on the willingness of the employee to undergo some degree of physical discomfort involved to achieve the protection. In cases such as this, the exact attenuation of noise exposure cannot be estimated exactly; however, its effectiveness can be measured audiometrically.

Now let's continue our simulated walk-around inspection on a construction site.

**Situation 22**

Perception: As we enter the site, we see two construction workers quietly talking.

Hazard: A much closer look reveals that one of the two "hard hats" is holding a portable electric radial saw at his side and that the saw has no guarding. Until the saw is put to use, no hazard exists unless it is accidentally dropped on the foot. When the saw is turned on, however, its operator will be immediately and unnecessarily exposed to the rapidly turning blade.

Abatement: The specially designed guard that came with this saw should be reattached before the saw is placed in use. A quick check of other saws and assorted power-operated equipment is worthwhile to identify whether such laxity is an isolated instance or a practice that pervades the entire construction site. This topic should be mentioned regularly as part of the safety training program.

**Situation 23**

Perception: At the top of the building under construction, an open elevator shaft is noted—eight floors up.

Hazard: Any employee or inspector could trip and fall to his or her death for lack of some physical barrier to keep people away from the shaft.

Abatement: Both a physical barrier and warning signs should be paced in proximity to this hazard. Employees should receive instructions regarding the necessity for isolating hazards such as this.

**Situation 24**

Perception: Walking through the storage and staging area, we note a dramatic example of a real problem that exists not only at construction sites, but in industrial plants, warehouses, and even retail operations—namely a stack of boxes that could be a model for the Leaning Tower of Pisa.

Hazard: The boxes are stacked too high, there are defective pallets, and the ground is so soft that inevitably the stack will come crashing down. We can only hope that no one will be near it when this occurs.

Abatement: Before anything else happens, the area surrounding the leaning stack of boxes should be posted as "off limits" to nonessential personnel. Then appropriate heavy equipment should be used to restack the boxes, using dependable pallets. When the emergency situation has been corrected, the long-term implications of such careless materials handling should be addressed.

As part of their safety training, personnel responsible for stacking operations should receive regular reminders of the importance of caution and care in the performance of their tasks. Other personnel, whose responsibilities take them to areas where materials are stacked, should be made aware of the necessity to stay alert to problems of this type and to report their existence right away.

**Situation 25**

Perception: We can't believe our eyes. A "hard hat" worker is walking directly beneath a load of plywood being transported by an overhead crane.

Hazard: Hard hat or no hard hat—the worst place a worker could possibly be is directly beneath an airborne load of anything, but this case is especially dangerous, since we see that only a single choker sling is in use.

Abatement: Assuming that the worker is unaware of the hazardous situation he has walked into, he should be warned to move away from the path of the overhead load. Following the abatement of the immediate threat to the worker, a series of cautionary remarks should be included as a part of the safety training program. Make sure that crane operators get the word too.

**Situation 26**

Perception: As we continue to observe the construction activities, we note two men operating air hammers while a third is clearing away broken pieces of concrete with a shovel. None wears eye, ear, or foot protection.

Hazard: None of the three has the necessary personal protection to ensure his individual safety and health. Excessive noise levels generated by the air hammers will, over a substantial exposure period, cause all three to lose some degree of their hearing capacity. All three are exposed to flying and falling fragments of the concrete as well, and they should have eye and foot protection.

Abatement: A program designed to increase desire to use personal protective equipment should be instituted. The desire is just as important as knowing how and when to use the equipment. No supervisor can keep an eye on all employees all the time, and if motivation to use the equipment does not exist, it will not be used or maintained properly.

**Situation 27**

Perception: Looking inside the mobile workshop, we see a bench grinder.

Hazard: There are no spark-deflecting guards on either the abrasive wheel or the wire wheel, and there is no face and eye protection visible. Additionally, the abrasive wheel is worn and needs to be replaced.

Abatement: Once the immediate hazards have been corrected, appropriate personnel should receive periodic reminders to be alert to the need to correct these hazards whenever they occur.

**Situation 28**

Perception: Also inside the mobile workshop is a floor-mounted buzz saw used to split lumber.

Hazard: As in Situation 22, personnel who operate the saw as it sets will be exposed to the dangers of an unguarded, rapidly spinning saw blade.

Abatement: The guard should be located and reinstalled on the saw before it is used again. A reminder sign should be posted to let saw operators know of the requirement to use the saw only when the guard is in place. Regular reference to this requirement should be included in safety training meetings.

**Situation 29**

Perception: Moving back outside we note that three levels of scaffolding are in use in constructing this brick-faced building.

Hazard: Numerous hazards exist in this situation. For example, inadequate cross-bracing on the uprights of the scaffolding, use of poor materials such as warped planks, and to top of this collection—there are no guard rails or toeboards. Just a moment. We note still another hazard—the center upright at the base of the wooden scaffolding is almost unbelievably insecure.

Abatement: Stop the entire operation and correct all the noted deficiencies before continuing the construction of this building. The person in charge of this site should be severely reprimanded or even dismissed, depending on his or her previous record. All future training should make note of the necessity to adhere to these and all safety-related requirements.

**Situation 30**

Perception: Passing by a generator used to power some electrical equipment, we stop to review the situation.

Hazard: The belt and pulley require guarding inasmuch as the starter button is so close that the operator must reach into the danger area to activate the generator.

Abatement: The guard for this part of the generator should be located and put into use prior to further operation. Personnel need to be cautioned not to forget to replace the guard after it has been removed for maintenance.

**Situation 31**

Perception: At the end of the site, we can see a long-necked crane (aerial boom) in operation.

Hazard: Of all the dangers involved in this type of operation, the use of this equipment in unreasonable proximity to overhead electrical lines is one of the most serious. Construction standards set forth specific operating clearances with respect to energized lines.

Abatement: As a regular part of the safety training program, all personnel working with or on cranes should be made aware of the possible results of operating a crane too close to loaded electrical lines. Furthermore, the company should require that only insulated, approved cabs be used when personnel are needed at the working end of a boom.

**Situation 32**

Perception: As we turn to leave the inspection site, a chain is spotted hanging on its storage rack.

Hazard: A closer look reminds us of the old story about a chain being only as strong as its weakest link. Here we see several links that appear to be worn and elongated—a characteristic of overloading. Also visible is evidence of improper splicing. So how strong is this chain?

Abatement: A regularly scheduled program of chain (and sling) inspection, maintenance, and replacement is needed. Personnel using chains should be trained to spot irregularities and to take appropriate action.

This view of some situations you and a compliance officer might see in the course of visiting workplaces was designed to provide insight into some safety and health hazards and their analysis. This is the type of hazard recognition training employees need to help them get started. It is never too late to begin.

## Suggested Inspection Inventory

Table 8.5 provides a suggested inspection inventory.

## Sample Policy for Employee Reports of Hazards

Every employee is expected to watch for possible hazards to employee safety and health and to report any noticed immediately. Reports may be made

**TABLE 8.5 Suggested Inspection Inventory to Help Guide the Development of Inspection Guidelines or Checklists**

1. Environmental factors (such as illumination, dusts, gases, sprays, vapors, fumes, noise, heat-cold)
2. Hazardous supplies and materials (such as explosives, flammables, acids, caustics, toxic materials or by-products)
3. Production and related equipment (such as mills, shapers, presses, borers, lathes, grinders, saws)
4. Power source equipment (such as steam and gas engines, electrical motors)
5. Electrical equipment (such as switches, fuses, breakers, outlets, cables, extension and fixture cords, ground connectors, connections)
6. Hand tools (such as wrenches, screwdrivers, hammers, chisels, power tools)
7. Personal protective equipment (such as hard hats, safety glasses, safety shoes, respirators)
8. Personal service and first-aid facilities (such as drinking fountains, wash basins, soap dispensers, safety showers, eyewash fountains, first aid supplies, stretchers)
9. Fire protection and extinguishing equipment (such as alarms, water tanks, sprinklers, standpipes, extinguishers, hydrants, hoses)
10. Walkways and roadways (such as ramps, docks, sidewalks, aisles, vehicle ways)
11. Elevators, electric stairways and manlifts
12. Working surfaces (such as ladders, scaffolds, catwalks, platforms, sling chairs, desk heights)
13. Material handling equipment (such as cranes, dollies, conveyors, hoists, forklifts, chains, ropes, slings)
14. Transportation equipment (automobiles, railroad cars, trucks, front-end loaders, helicopters, motorized carts and buggies)
15. Warning and signaling devices (such as sirens, crossing and blinker lights, claxons, warning signs)
16. Containers (such as scrap bins, disposal receptacles, carboys, barrels, drums, gas cylinders, solvent cans)
17. Storage facilities and areas, both indoor and outdoor (such as bins, racks, lockers, cabinets, shelves, tanks, closets)
18. Structural openings (such as windows, doors, stairways, sumps, shafts, pits, floor openings)
19. Buildings and structures (such as floors, roofs, walls, fencing)
20. Miscellaneous

*Source:* National Safety Council/OSHA Training on Safety and Health Management.

orally to the employee's supervisor or through the written reports system (use your own name for the system here) immediately or as soon as possible.

No employee, at any level, shall discipline or harass any other employee because of reports of hazards made. Any employee found to have discriminated against another employee for this reason shall be disciplined.

Remember, *you* are needed to help keep this worksite safe and healthful!

(Signature of highest level of management at worksite.)

**Example 1: Employee Report of Hazard**

EMPLOYEE: ______________________________
HOUR: ______________
DATE: ______________
DEPARTMENT WHERE HAZARD OBSERVED: ______________
HAZARD OR PROBLEM: ______________________
______________________________________________
______________________________________________
SUGGESTED ACTION TO TAKE: __________________
______________________________________________
______________________________________________
______________________________________________

EMPLOYEE: COMPLETE AND GIVE TO SUPERVISOR

______________________________________________
______________________________________________
SUPERVISOR: ______________________________
DEPARTMENT: ______________________________
DATE: ______________
ACTION TAKEN: ____________________________
______________________________________________
______________________________________________
______________________________________________

SUPERVISOR: COMPLETE AND GIVE TO MANAGER

______________________________________________
______________________________________________
DATE: ______________
REVIEW - COMMENTS: ________________________
______________________________________________

______________________________
Signature of Manager

***Follow-up Documentation***

(Can be used as part of the preceding form or separately in the case where employees are not required to put the report in writing.)

HAZARD: ______________________________

______________________________

Possible Injury or Illness: ______________________________

______________________________

Exposure: ______________ Frequency: ______________
Duration: ______________________________
Interim Protection Provided: ______________________________

______________________________

Corrective Action Taken: ______________________________

______________________________

______________________________

______________________________

______________________________

______________________________

______________________________

Follow-up Check Made on ______________. Any additional action taken?
(date)

______________________________

______________________________

______________________________
Signature of Manager or Supervisor

______________________________
Date

Three month follow-up check made on ______________________________.
Date

Is corrective action still in place? ______________ Yes ______________ No

**Example 2: Report of Safety or Health Problem**

1. Date ______________
2. Description of Problem (including, if applicable, exact location) ______

______________________________

______________________________

______________________________

3. Note any previous attempt to notify on this problem and to whom

______________________________

______________________________

Date ______________
Submitted by (optional) Name ______________________________

______________________________

Safety Department Findings: ______________________________

______________________________

______________________________

Action Taken ____________________________________________
____________________________________________
____________________________________________

Safety Committee Review ____________________________________________
Comments ____________________________________________
____________________________________________
____________________________________________

*All* actions *completed* by (date) ______________________________

### Example 3: Employee Report of Hazard(s)

**Note: This form is provided for the assistance of the employee making the report and does not constitute the exclusive means for notification of hazards.**

I believe that a condition or practice at the following location is a job safety or health hazard.

Is there an immediate threat of death or serious physical harm?
Yes __________ No __________

Provide information that will help locate the hazard(s) such as building or area of building or the supervisor's name. ______________________________
____________________________________________
____________________________________________

Describe briefly the hazard that you believe to exist and approximately the number of employees exposed to it. ______________________________
____________________________________________
____________________________________________
____________________________________________

If this hazard has been called to anyone's attention, as far as you know, please provide the name of the person or committee notified and the approximate date. ____________________________________________
____________________________________________
____________________________________________

Signature (optional) ______________________________
Typed or printed name (optional) ______________________________
Date ____________________
Management evaluation of reported hazard(s) ______________________________
____________________________________________
____________________________________________

Final action taken ______________________________
____________________________________________
____________________________________________
____________________________________________

All actions completed by ______________ (date) Initials ______________

Figure 8.5 presents hazardous analysis flow charts.

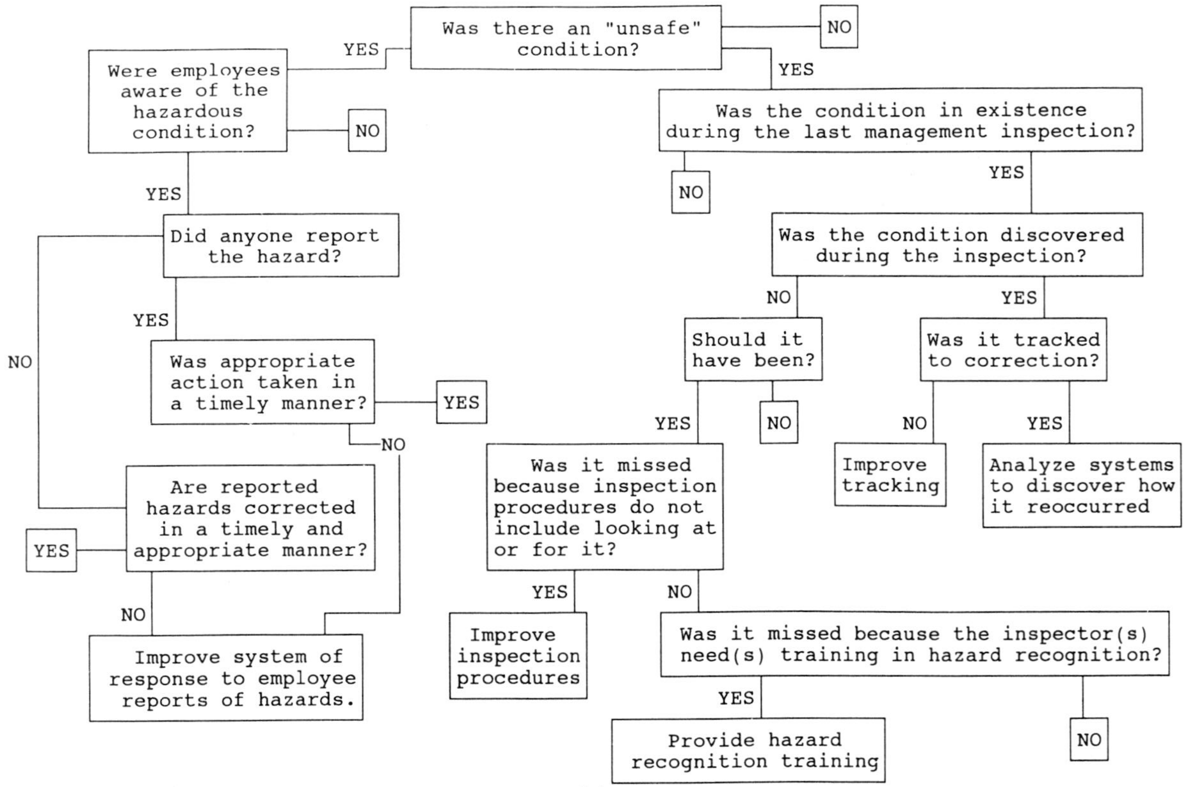


(*a*)

**FIGURE 8.5** Hazard analysis flow charts.

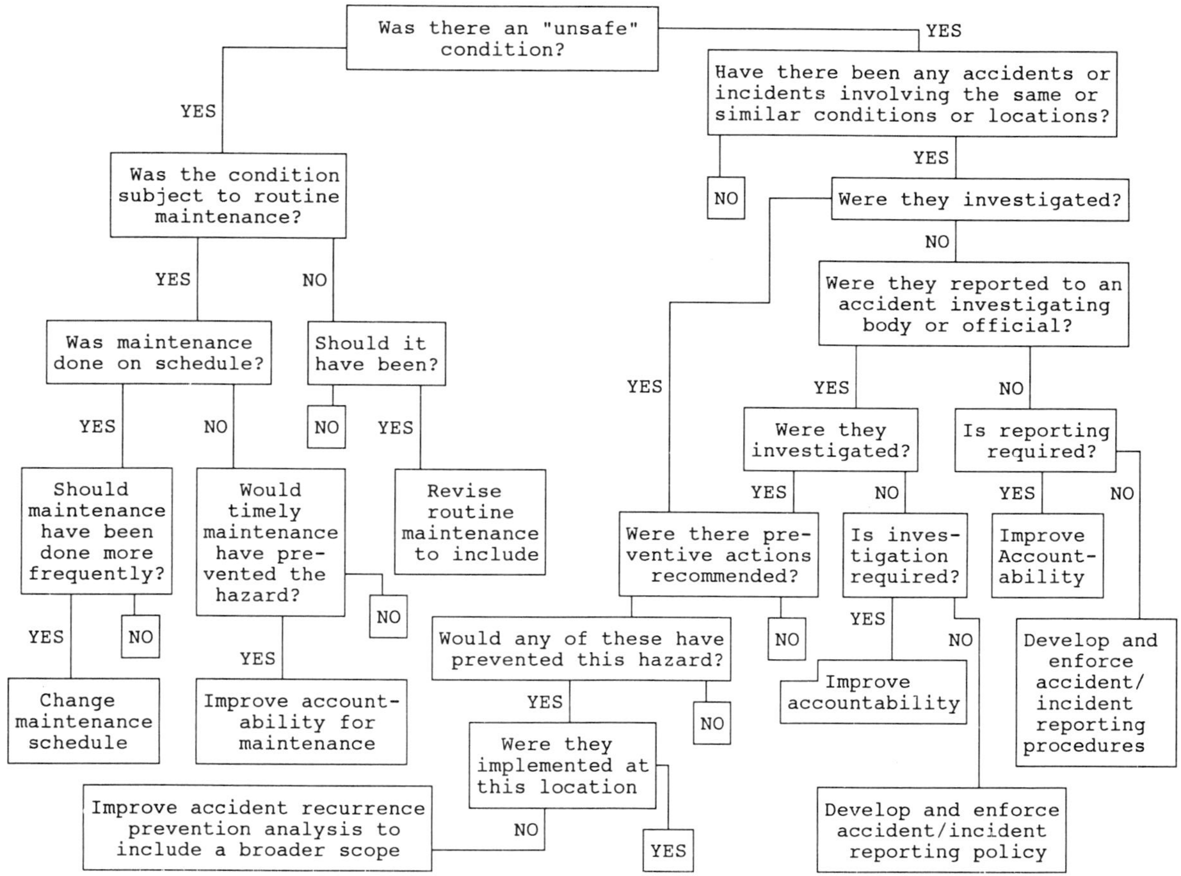


(b)

**FIGURE 8.5** *(Continued)*

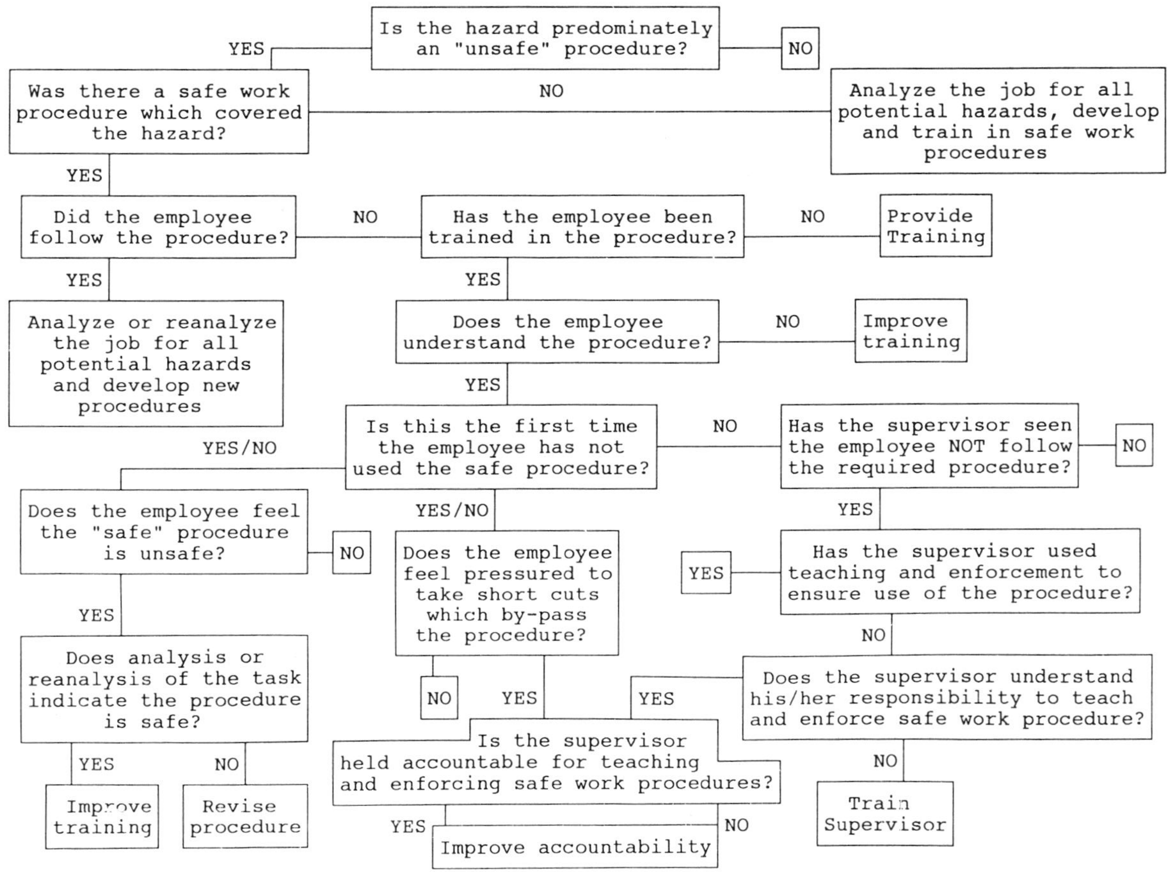


(c)

**FIGURE 8.5** *(Continued)*

## OSHA AND OSHA INSPECTION TRAINING

### Background

The OSHAct includes the following requirements:

Each employer shall furnish to his or her employees employment and a place of employment that are free from recognized hazards that are causing or are likely to cause death or serious physical harm to his or her employees.

Each employer shall comply with the occupational safety and health standards promulgated under this Act. See Occupational Safety Health Standards.

Each employee shall comply with occupational safety and health standards and all rules, regulations, and orders issued pursuant to this Act that are applicable to his or her own actions and conduct.

Each employer and employee shall comply with the training requirements contained in Part 1910 of the Code of Federal Regulations (CFR), commonly referred to as "General Industry" standards.

#### *OSHA Inspections*

The Department of Labor has hundreds of trained and experienced inspectors who are called safety and health compliance officers (COs). The OSHA Cos will visit employers for three reasons:

- Death of an employee from work-related causes
- Routine industry inspections in an attempt to improve the working conditions of the employees in that industry
- Complaints from an employee, if they are in writing, and of sufficient merit to cause OSHA to act

#### *Compliance Officer's Scope*

The OSHAct provides the compliance officer with the following authority and responsibility:

- Full access at all reasonable hours
- Must explain why he or she is there
- May request accompaniment on tour by employee representative or employer's representative
- Is in charge of inspection tour
- May talk with any employee
- Can measure or test noise, air, dust, heat, chemicals, and so on
- May photograph area or location
- Will examine records
- Can request and copy specific medical records and any monitoring data

- Can seek identity of any substances suspected of being dangerous or hazardous; may also take a sample of suspect substance
- May request shutdown, discontinuance of operations, removal of substances, or removal of personnel only when he or she decides there is an imminent danger situation
- Is expected to take proper tests and careful notes
- Must discuss his or her findings with management prior to leaving
- Will advise any complainant whether there was actually a violation
- Can issue a citation for violation
- May recommend a penalty assessment

### *Supreme Court Rulings*

On May 23, 1978, the United States Supreme Court ruled in the Barlow case that businesses could refuse warrantless OSHA inspections. According to the Research Institute of America, this decision by the Court put several heretofore unavailable options in the hands of management.

Refusing a warrantless inspection not only forces the OSHA inspector to obtain a warrant, it also limits the scope of his or her search. To obtain that warrant, OSHA must specify what it is looking for. That rules out "wall-to-wall" inspections without probable cause.

Despite the Court's landmark ruling, only a very small percentage of the employers visited by OSHA's compliance officers since the Barlow decision have demanded warrants prior to permitting entrance.

Some recommendations for preparing the workforce for an OSHA inspection include

- Advising everyone to use a tight-lipped, but cooperative approach, that is, answer all questions honestly, but do not volunteer information not specifically requested
- Before the inspector departs, be sure to ask him or her what areas might receive citations, why, and how to abate them.

Before deciding to appeal (3 weeks from the date of the receipt of the OSHA citation), a firm should request an informal conference with the area director. Sometimes in this way differences can be resolved without OSHA issuing a formal complaint.

Any appeal should explain why the company believes the citation is either incorrect or unreasonable (e.g., unreasonable abatement times, excessive fines, unavailable technology).

Any appeal should represent an affirmative defense: a faulty search warrant, impossibility of compliance (without reference to cost), unfeasability of controls, or perhaps proof that the cited violation was an isolated event.

Paying fines just to avoid the expensive and time-consuming legal process can backfire and open a company up to possible civil liability suits.

OSHA Directive 200-82 instructs its field offices to afford union or employee representatives the opportunity to participate in the opening and closing conferences of an inspection and in any informal conferences between OSHA and businesses where disputes are discussed. The directive also states that any withdrawn or amended violation can be appealed by the union or employee representative to one of the agency's 10 regional offices.

## Training Responsibilities

Two broad provisions of the OSHAct are applicable to every employer:

Every employer must be able to produce records indicating that each employee has received training in the areas identified in the standards. Every employer must be able to show that each employee has received training based on an analysis of tasks performed.

It is easily understood that employers must be able "to produce records" regarding employee training, but what is meant by the term "in the areas identified in the standards?"

OSHA 2254 specifically sets forth all the sections of the act relating to training requirements. These are the "areas identified in the standards." A review of these standards shows them, for the most part, to be directed to particularly hazardous jobs or industries.

## Hazard Abatement and Appeal Procedures

Citations must be posted near the site of violation. Posting is to be made on the day of receipt and must remain for 3 days or until the hazard has been abated, whichever is longer. Notices of appeal may also be posted. Abatement of a cited condition prior to 3 days means a notice must be posted alongside the citation.

Appeals from citations and proposed penalties may be made, but a citation becomes binding on employers if not contested within 15 working days of receipt. If the employer wishes to contest a citation, he or she must base the appeal on one of the following points:

- The citation itself
- The proposed penalty
- The period established by abatement
- Or any combination of these

## Imminent Danger

OSHAct defines "imminent danger" as any condition that would be expected to cause serious injury or death.

When a complaint is made that an imminent danger exists, the compliance officer may make a site investigation. If the condition does in fact present an imminent danger, the compliance officer will do the following:

- Post a citation
- Notify employer
- Notify employees affected
- Seek a court order to abate the violation. This could be a request to shut down the equipment or stop using the material in question, and/or remove all personnel from the area except those needed to abate the condition.

## SUMMARY

Hazard recognition and control must be a cornerstone of any safety and health program. Logically, safety and health training must provide the entire workforce with the knowledge and skills to perform effective inspections to identify hazards and potential hazards and to develop programs to control these hazards. Hazard recognition and inspection training should thus be a very significant element in every safety and health training program.

## CHAPTER 9

# HAZARD COMMUNICATIONS

## INTRODUCTION

One of the more recent and, for many organizations, difficult to implement OSHA requirements is the requirement for a hazard communications program under the provisions of 29 CFR 1910.1200.

This chapter will provide general information on the required hazard communication program elements and specific information on hazard communication training.

The purpose of the hazard communication program is to ensure that hazards associated with chemicals produced in the United States or imported are evaluated and hazard information, including appropriate protective, cleanup, and disposal methods, are communicated to those who work with and around these chemicals.

## HAZARD COMMUNICATION PROGRAM ELEMENTS

There are five major components to the hazard communication standard (Figure 9.1). These are the building blocks of the effort to ensure that persons know and understand how to handle and store hazardous substances.

### Hazard Determination

The first component is hazard determination. Manufacturers and importers have the primary responsibility for determining and evaluating the physical

FIVE MAJOR COMPONENTS
OF HAZCOM STANDARD

- HAZARD DETERMINATION
  by manufacturers/importers
- WRITTEN HAZCOM PROGRAM
- LABELS
- MATERIAL SAFETY DATA SHEETS
- INFORMATION AND TRAINING

**FIGURE 9.1** Five major components of a hazard communications program.

and health hazards associated with chemicals that they manufacturer or import.

Employers or users of chemicals are not required to determine or evaluate hazards, but may do so if they do not have confidence in the material supplied by manufacturers/importers.

### Written Hazard Communication Program

Employers and users of chemicals subject to OSHA must produce a written hazard communication program that outlines how the company plans to comply with the provisions of 29 CFR 1910.1200.

The written program should specify how hazardous chemicals will be listed in each work area, details about container labeling, the process for obtaining, handling, distributing, storing, and updating material safety data sheets, and the methods of providing employee information and training.

### Labels

The third major element of the hazard communication standard specifies requirements for labeling containers of hazardous chemicals (Figure 9.2).

The OSHA standard states that a container label must contain the following:

- The name and address of the manufacturer
- The identity of the chemical inside
- Hazard warnings for employee protection

LABELS

- NAME AND ADDRESS OF MANUFACTURER
- IDENTITY OF CHEMICAL
- HAZARD WARNINGS FOR EMPLOYEE PROTECTION

**FIGURE 9.2** Labels

Manufacturers/importers are responsible for ensuring that all containers are properly labeled prior to leaving the workplace.

Employers are responsible for ensuring that all containers received, stored, and used in the workplace are properly labeled and that labels are not removed or defaced.

MATERIAL SAFETY DATA SHEET (MSDS)

- SECTION I - CHEMICAL ID
- SECTION II - HAZARDOUS INGREDIENTS
- SECTION III - PHYSICAL DATA
- SECTION IV - FIRE AND EXPLOSION HAZARD DATA
- SECTION V - HEALTH HAZARD DATA
- SECTION VI - REACTIVITY DATA
- SECTION VII - SPILL/LEAK PROCEDURES
- SECTION VIII - SPECTION PROTECTION INFORMATION
- SECTION IX - SPECIAL PRECAUTIONS

**FIGURE 9.3** Material safety data sheets.

OSHA does not specify a specific type of label. A variety of labels are acceptable as long as they contain the required information.

## Material Safety Data Sheets

The manufacturer/importer of hazardous materials is required to develop a material safety data sheet (MSDS) for each hazardous chemical produced or imported.

These sheets contain detailed information about the products and must be passed on to distributors and to end users. They supplement label information and provide the detailed information required to safely handle, store, and use chemical products. The MSDS also provide information necessary to safely respond to spills, leaks, or other emergencies.

Each MSDS should include nine sections:

- Section I contains the chemical identification of the chemical, the name and address of the manufacturer/importer, and an emergency telephone number.
- Section II describes the hazardous ingredients in the product, including the percent by volume and the threshold limit value (TLV) of each ingredient.
- Section III provides the physical data about the product, including the boiling point, vapor pressure, vapor density, solubility in water, specific gravity, percent volatile by volume, evaporation rate, appearance, and odor.
- Section IV contains fire and explosion hazard data, including the flash point, flammable limits, extinguishing media, special fire fighting procedures, and unusual fire and explosion hazards.
- Section V details health hazard data. This includes the threshold limit value (TLV), the effects of overexposure, and the emergency first aid procedures.
- Section VI provides reactivity data, including stability, incompatibility, hazardous decomposition products, and hazardous polymerization information.
- Section VII lists spill or leak procedures, including the steps to be taken in case of release or spill and acceptable and prohibited waste disposal methods.
- Section VIII outlines special protection information, such as the need for respiratory protection, ventilation requirements, the type of glove to use (if needed), and the type of eye protection needed, including conditions of use or limitations.
- Section IX discusses special precautions, including handling and storage precautions and any other precautions not already listed on the MSDS.

### Employee Information and Training

The fifth element of the program consists of the information and training that must be provided to employees. This element is described in detail below.

## HAZARD COMMUNICATION TRAINING

The following is extracted from 29 CFR 1910.1200:

> (h) Employee information and training. Employers shall provide employees with information and training on hazardous chemicals in their work area at the time of their initial assignment, and whenever a new hazard is introduced into their work area.
>
> (1) Information. Employees shall be informed of:
>
> (i) The requirements of this section;
>
> (ii) Any operations in their work area where hazardous chemicals are present; and,
>
> (iii) The location and availability of the written hazard communication program, including the required list(s) of hazardous chemicals, and material safety data sheets required by this section.
>
> (2) Training. Employee training shall include at least:
>
> (i) Methods and observations that may be used to detect the presence or release of a hazardous chemical in the work area (such as monitoring conducted by the employer, continuous monitoring devices, visual appearance or odor of hazardous chemicals when being released, etc.);
>
> (ii) The physical and health hazards of the chemicals in the work area;
>
> (iii) The measures employees can take to protect themselves from these hazards, including specific procedures the employer has implemented to protect employees from exposure to hazardous chemicals, such as appropriate work practices, emergency procedures, and personal protective equipment to be used; and,
>
> (iv) The details of the hazard communication program developed by the employer, including an explanation of the labeling system and the material safety data sheet, and how employees can obtain and use the appropriate hazard information.

## SUMMARY

To meet these requirements, it may be necessary, especially in larger and/or diverse organizations, to decentralize at least part of the hazard communications training. Such things as the details of the employer's hazard communication program, explanations of labeling, and general information about how to use and read material safety data sheets may be taught by a central training organization since this information is common to all employees, but frequently the specifics of the hazards and chemicals associated with individuals processes, facilities, or tasks are best communicated by supervisors. This job-specific information is critical to an effective program. Ensuring that this information is provided in compliance with the OSHA requirement requires effective monitoring and documenting if a decentralized approach is used. This can be very challenging in large, diverse organizations, especially those using a number of different hazardous chemicals.

## CHAPTER 10

# LOCKOUT AND TAGOUT

### INTRODUCTION

While lockout and tagout programs and requirements have been around for many years, 29 CFR 1910.147, "The Control of Hazardous Energy" (January 1990), greatly expanded the requirements for lockout/tagout programs and training.

These new requirements apply to most general industry operations, but the standard does *not* cover construction, agriculture, maritime operations, oil and gas well drilling and servicing, installations under the exclusive control of electrical utilities for the purpose of power generation, transmission and distribution, or exposure to electrical hazards, which are covered by subpart S of 29 CFR 1910.

This chapter will outline 29 CFR 1910.147 requirements for tagout/lockout programs and the specific training requirements for the program.

An example of a tagout/lockout procedure, contained in 29 CFR 1910.147, is also included.

### OSHA TAG AND LOCKOUT REQUIREMENTS

The employer is required to establish an energy control program consisting of procedures, training, and periodic inspections to ensure that before a machine is serviced or repaired, any potentially harmful energy source associated with the machine is isolated and rendered inoperable (see 29 CFR 1910.147 paragraph c).

Energy isolating devices (switches, valves, etc.) must be locked out or tagged out. If possible the device must be locked out unless the employer can demonstrate that a tagout system will provide full employee protection. Unless the energy isolating devices are incapable of being locked out (equipment replaced, renovated, or modified after January 2, 1990, must be designed to accept a lockout device), it is strongly recommended that locks be used.

Those who really believe that tagouts can offer the same protection as lockouts should demonstrate their commitment to this principle by hanging a "DO NOT OPERATE" tag on the steering wheel of their cars and on the door knobs of their houses and leave their cars and houses unlocked. Additionally, to convince OSHA inspectors (or any other rational creatures) that a tagout system provides equivalent safety is extremely difficult. Procedures have to be more detailed and training and enforcement efforts must be clearly defined and well documented.

The lockout/tagout procedure is an essential element of the lockout/tagout program. Good procedures are also essential for good training, since the training program must be directly derived from the procedure.

The following are the procedure requirements as stated in 29 CFR 1910.147:

> The procedures shall clearly and specifically outline the scope, purpose, authorization, rules, and techniques to be utilized for the control of hazardous energy, and the means to enforce compliance including, but not limited to, the following:
>
> (A) A specific statement of the intended use of the procedure;
>
> (B) Specific procedural steps for shutting down, isolating, blocking and securing machines or equipment to control hazardous energy;
>
> (C) Specific procedural steps for the placement, removal and transfer of LOCKOUT devices or TAGOUT devices and the responsibility for them; and
>
> (D) Specific requirements for testing a machine or equipment to determine and verify the effectiveness of the LOCKOUT devices, TAGOUT devices, and other energy control measures.

The OSHA regulation on lockout/tagout contains a sample "Typical Minimal LOCKOUT Procedure." It should be noted that this is a guide only. For more complex systems and/or if tagout is used, more detailed and more comprehensive procedures are required.

The sample procedure from Appendix A to 1910.147 follows:

LOCKOUT PROCEDURE

LOCKOUT procedure for ____________________________

(Name of Company for single procedure or identification of equipment if multiple procedures are used)

### Purpose

This procedure establishes the minimum requirements for the LOCKOUT of energy isolating devices whenever maintenance or servicing is done on machines or equipment. It shall be used to ensure that the machine or equipment is stopped, isolated from all potentially hazardous energy sources and locked out before employees perform any servicing or maintenance where the unexpected energization or start-up of the machine or equipment or release of stored energy could cause injury.

### Compliance With This Program

All employees are required to comply with the restrictions and limitations imposed upon them during the use of LOCKOUT. The authorized employees are required to perform the LOCKOUT in accordance with this procedure. All employees, upon observing a machine or piece of equipment which is locked out to perform servicing or maintenance shall not attempt to start, energize or use that machine or equipment.

______________________________________________

(Type of compliance enforcement to be taken for violation of the above)

### Sequence of LOCKOUT

(1) Notify all affected employees that servicing or maintenance is required on a machine or equipment and that the machine or equipment must be shut down and locked out to perform the servicing or maintenance.

______________________________________________

(Names/Job Titles of affected employees and how to notify)

(2) The authorized employee shall refer to the company procedure to identify the type and magnitude of the energy that the machine or equipment utilizes, shall understand the hazards of the energy, and shall know the methods to control the energy.

______________________________________________

(Types and magnitudes of energy, its hazards and the methods to control the energy)

(3) If the machine or equipment is operating, shut it down by the normal stopping procedure (depress stop button, open switch, close valve, etc.).

______________________________________________

(Types and locations of machine or equipment operating controls)

(4) De-activate the energy isolating device(s) so that the machine or equipment is isolated from the energy source(s).

______________________________________________

(Types and locations of energy isolating devices)

(5) Lock out the energy isolating device(s) with assigned individual lock(s).

(6) Stored or residual energy (such as that in capacitors, springs, elevated machine members, rotating flywheels, hydraulic systems, and air, gas, steam, or water pressure, etc.) must be dissipated or restrained by methods such as grounding, repositioning, blocking, bleeding down, etc.

______________________________________________

(Types of stored energy—methods to dissipate or restrain)

(7) Ensure that the equipment is disconnected from the energy source(s) by first checking that no personnel are exposed, then verify the isolation of the equipment by operating the push button or other normal operating control(s) or by testing to make certain the equipment will not operate.

Caution: Return operating control(s) to neutral or "off" position after verifying the isolation of the equipment.

---

(Method of verifying the isolation of the equipment)

(8) The machine or equipment is now locked out.

Restoring Equipment to Service

When the servicing or maintenance is completed and the machine or equipment is ready to return to normal operating condition, the following steps shall be taken:

(1) Check the machine or equipment and the immediate area around the machine or equipment to ensure that nonessential items have been removed and that the machine or equipment components are operationally intact.

(2) Check the work area to ensure that all employees have been safely positioned or removed from the area.

(3) Verify that the controls are in neutral.

(4) Remove the LOCKOUT devices and reenergize the machine or equipment.

Note: The removal of some forms of blocking may require reenergization of the machine before safe removal.

(5) Notify affected employees that the servicing or maintenance is completed and the machine or equipment is ready for use.

## TAGOUT AND LOCKOUT TRAINING

Under the provisions of 29 CFR 1910.147, training is required for both "authorized employees" and "affected employees."

Definitions, as listed in 1910.147, are as follows:

Authorized Employee: A person who locks out or tags out machines or equipment in order to perform servicing or maintenance on that machine or equipment. An affected employee becomes an authorized employee when that employee's duties include performing servicing or maintenance covered under this section.

Affected Employee: An employee whose job requires him/her to operate or use a machine or equipment on which servicing or maintenance is being performed under LOCKOUT or TAGOUT, or whose job requires him/her to work in an area in which such servicing or maintenance is being performed.

Generally, it is recommended that all service/maintenance organization employees (and all other employees who perform service/maintenance tasks) be trained as authorized employees and that all other employees be trained as affected employees.

Training requirements, as listed in 29 CFR 1910.147, are as follows:

Training and communication:

The employer shall provide training to ensure that the purpose and function of the energy control program are understood by employees and that the knowledge and skills required for the safe application, usage, and removal of the energy controls are acquired by employees. The training shall include the following:

(A) Each authorized employee shall receive training in the recognition of applicable hazardous energy sources, the type and magnitude of the energy available in the workplace, and the methods and means necessary for energy isolation and control.

(B) Each affected employee shall be instructed in the purpose and use of the energy control procedure.

(C) All other employees whose work operations are or may be in an area where energy control procedures may be utilized, shall be instructed about the procedure, and about the prohibition relating to attempts to restart or re-energize machines or equipment which are locked out or tagged out.

When TAGOUT systems are used, employees shall also be trained in the following limitations of tags:

(A) Tags are essentially warning devices affixed to energy isolating devices, and do not provide the physical restraint on those devices that is provided by a lock.

(B) When a tag is attached to an energy isolating means, it is not to be removed without authorization of the authorized person responsible for it, and it is never to be bypassed, ignored, or otherwise defeated.

(C) Tags must be legible and understandable by all authorized employees, affected employees, and all other employees whose work operations are or may be in the area, in order to be effective.

(D) Tags and their means of attachment must be made of materials which will withstand the environmental conditions encountered in the workplace.

(E) Tags may evoke a false sense of security, and their meaning needs to be understood as part of the overall energy control program.

(F) Tags must be securely attached to energy isolating devices so that they cannot be inadvertently or accidentally detached during use.

Employee retraining.

(A) Retraining shall be provided for all authorized and affected employees whenever there is a change in their job assignments, a change in machines, equipment or processes that present a new hazard, or when there is a change in the energy control procedures.

(B) Additional retraining shall also be conducted whenever a periodic inspection under paragraph (c)(6) of this section reveals, or whenever the employer has reason to believe, that there are deviations from or inadequacies in the employee's knowledge or use of the energy control procedures.

(C) The retraining shall reestablish employee proficiency and introduce new or revised control methods and procedures, as necessary.

The employer shall certify that employee training has been accomplished and is being kept up to date. The certification shall contain each employee's name and dates of training.

## SUMMARY

An effective LOCKOUT/TAGOUT program will include detailed procedures for each machine or piece of equipment which has the potential to injure an employee during service or maintenance operations.

Authorized employees should be trained, and retrained as changes occur, on all aspects of these procedures.

Affected employees should be trained, and retrained as changes occur, on the purpose of the procedures and the requirements to respect locks and tags used to render machines or equipment safe during service or maintenance operations.

# CHAPTER 11

# FIRST AID/CPR TRAINING AND MEDICAL OPERATIONS

## INTRODUCTION

To protect the safety and health of the workforce, access to emergency medical treatment and other medical services is essential.

The nature of the emergency treatment and other medical services varies widely depending upon several factors, including the nature of the work, the location of the job site, and the size of the organization.

For example, many construction sites and small businesses in remote areas frequently have to rely very heavily on first aid/CPR-trained members of the workforce to provide emergency response.

Large organizations may have in-house medical staffs that can provide emergency and other medical services.

Businesses and job sites located near private or public hospitals may rely primarily on the services provided by the hospitals.

This chapter discusses first aid/CPR training requirements. It also discusses medical programs and includes OSHA pamphlet 11, "The Occupational Health Delivery System: A Medical Program that is Right for Your Worksite," produced by the OSHA Office of Consultation Programs.

## MEDICAL AND HEALTH PROGRAM OBJECTIVES

The health and safety of employees must receive top priority in every business. Because professional medical help is seldom immediately available, it is imperative that emergency treatment be available when an injury occurs.

Occupational health programs are concerned with all aspects of a worker's health. The basic objectives of such a program, according to the American Medical Association, should be the following:

- To protect employees against health hazards in their work environment
- To facilitate job placement and to ensure the suitability of individuals according to their physical capacities, mental abilities, and emotional makeup without danger to their health and safety or to that of their fellow employees
- To ensure adequate medical care and rehabilitation of those injured or made ill on the job
- To encourage personal health

The American Medical Association has listed the following elements and services that should be provided to maintain the health of every work force, regardless of size of the operation:

- Maintenance of a healthful environment
- Diagnosis and treatment
- Health examinations
- Immunization programs
- Medical records
- Health education and counseling
- Free communication between the plant physician and the personal physician

Achieving these goals benefits both employee and employer. If employees are to perform the jobs safely and efficiently, they must be in good health.

## Personnel Selection and Training

Assigning the right person to the right job and protecting him or her against the hazards of the work are, however, only the beginning. Medical personnel today must be familiar with the on-the-job hazards, and they must continually assess the job hazards just as they periodically reexamine employees to ensure their physical fitness.

Working as a team, the medical and safety personnel have to consider the whole person in the total environment—not just on the job.

Programs to rehabilitate alcoholic employees, those with a drug problem, and so on, are becoming more common in plants that at one time simply dismissed an employee who was found to be addicted to alcohol or other drugs.

Some argue that such plants are just going "soft," becoming overindulgent, and inviting trouble. It is true that in the past management took a sterner view;

but life was less complicated then, and we understood less about the etiology of these and other problems. Today, every employee is an investment, and it is usually to management's benefit to protect that investment.

### First Aid and Cardiopulmonary Resuscitation (CPR)

First aid can be defined as temporary aid and comfort to avert further complications in an emergency until professional medical attention can be obtained. First aid has never been intended to take the place of a physician, nurse, or paramedic. Rather, it should be an integral part of the complete safety program of a plant. First aid programs should include CPR training.

Plants should appoint a safety officer to be responsible for the entire first aid program. This person, in turn, may enlist supervisory personnel for assistance. It is important, however, to have one individual coordinate the program.

Each department, shift, or work group should have at least one person trained in first aid and CPR. The number of trained individuals and the extent of their training will depend on occupational hazards, location of operations in relation to immediate medical attention, and the size of the work unit.

The safety officer should coordinate all first aid training. Free or low cost instruction is available from the American Red Cross and some state labor departments. The Red Cross offers several different courses of instruction in first aid and CPR, ranging from a very basic course for those who have difficulty with the English language, to advanced courses for first aid instructors. Such trained individuals have proved invaluable to many plant first aid programs.

### Training

All employees should be familiar with basic first aid and CPR techniques. Supervisors and foremen should have more extensive first aid training. One method of developing the first aid and CPR skills of supervisors is to have them attend first aid and CPR "Train the Trainer" courses and conduct first aid and CPR training to their employees.

To assist in meeting the need for personnel trained in first aid as required by the OSHAct regulations, the Secretary of Labor and the president of the American National Red Cross have agreed on the following points:

- Department of Labor recognizes persons who have current Red Cross certificates as fulfilling the regulation.
- Red Cross will develop guidelines to expand first aid training to meet the needs of different work environments.
- Red Cross will distribute information relating to first aid training services available to employers and employees.

**Note:** OSHA has entered into an agreement with the American National Red Cross giving OSHA approval to Red Cross first aid training and calling for close cooperative efforts between OSHA and the Red Cross.

There is no requirement that personnel be trained according to Red Cross methods, but Red Cross training has received specific OSHA approval.

Whatever the source of first aid/CPR training, it is essential that the training include warnings about the dangers of bloodborne pathogens, especially the hepatitis B virus and acquired immune deficiency syndrome (AIDS) virus. This training is now required by OSHA for all health providers and others that may be at risk.

Refresher courses should be taken at least once a year because of changing medical practices, and because company employee rosters are subject to personnel changes that can disrupt the first aid program within a plant or work group. The safety officer should regularly review the list of employees with first aid training certificates.

Educating all employees about the first aid program is essential. They can be kept informed through bulletin boards, newsletters, posters, and so on. Location of first aid kits and the names of employees trained in first aid and CPR procedures should be well publicized.

### Medical Records and Facilities

The occupational nurse, or the employee in charge of the dispensary, is usually assigned the task of maintaining employees' medical records. These records are kept according to standard procedures, and the job should not be so laborious that it requires most of the time of the person so assigned. Of course, the volume may be so great that recordkeeping is the sole responsibility of the person.

Although medical records are confidential, the data should be recorded in standardized language and symbols to be understood by anyone so trained. Standardization also facilitates reliable statistical comparisons and analysis.

To minimize ambiguity and to ensure comparability of terms, records can be made in the terminology of the American Medical Association's *Standard Nomenclature of Diseases and Operations.* The AMA has another useful publication: *Guide to Development of an Industrial Medical Records System.* (For the address of the AMA, see Appendix "Service Organizations and Association.")

The unusual recommendation for a dispensary suggests a minimum of three rooms, a waiting room, a treatment room, and a room for consultation or for making physical examinations. A properly outfitted examining room can serve also for audiometric testing and other purposes. The treatment room should be large enough to treat more than one person at a time.

It is advisable to also have at least one first aid room or station at a location that is convenient to the employees. A room for administering first aid is usually more acceptable, because of the need for privacy.

The first aid room should have an examining table, a cot, a dustproof cabinet for supplies, and a fully equipped first aid kit, a covered waste receptacle, a magnifier with a light, a wheelchair, a stretcher, blankets, a phone with important numbers posted for emergencies, and one or two chairs and a table.

A major step in setting up a first aid program is to select the proper first aid kit. There are many types, designed to fill varying needs, depending on the hazards that might occur. One of the most useful is the unit-type kit. Relatively inexpensive, it is quite suitable for use when nurses are not nearby. The kits have standard-size packages, each containing a single type of dressing or treatment—such as burn cream, ammonia inhalant, eyewash solution, bandage compresses, or aspirin tablets. It is very important that first aid kits contain rubber gloves, breathing tubes, and other supplies to protect first aiders from the blood and bodily fluids of patients.

No measuring or sorting is necessary, minimizing time and effort when treatment is needed. The unit-type kits are available from several manufacturers and can be furnished to hold 6, 10, 16, 24, or 36 units. Other advantages of the kits are as follows:

- Each unit carton in the kit is overwrapped in a see-through protective covering to keep the contents clean and to indicate whether the carton has been opened.
- Each unit carton includes instructions and/or diagrams to help persons use the contents.
- All unit-type kit manufacturers offer similar materials packaged in identical size cartons. Restocking from different manufacturers causes no inconvenience.
- All contents are immediately visible and easily identified when the kit is opened.
- Company logos can be placed on the kits with decals or by silk screening.
- Kits are weatherproof and made of steel or high-impact plastic. Each kit has a carrying handle and hangers for wall mounting.

Since OSHA regulations require that all first aid supplies be approved by a consulting industrial physician, it may be helpful to furnish this individual with a complete list of materials available so that he or she can recommend the best combination. Specific requirements are determined by the number of employees and the hazards to which they may be exposed. Once the physician has selected the contents, a certificate of approval should be obtained from him or her and filed for possible future inspections by OSHA compliance officers.

Another key action in instituting a first aid program is determining where to place the first aid kits. They should be located so that no worker in the plant is more than a few minutes away from first aid treatment. Self-evaluation checklists available from the National Safety Council can be helpful in this determination.

OSHA requires that the color green be used to identify each area designated for first aid, but this rule has been misinterpreted in many ways. Some plants, for example, have painted their first aid kits green. Kits can be any color, but the immediate area in which they are located must be green.

Telephone numbers of selected doctors, hospitals, and ambulances should be placed next to the kits. Such information is particularly helpful to persons working on the night shifts.

A depleted kit is of no use and is an OSHA violation. The safety officer has the responsibility to keep each kit well stocked. Unfortunately, a common problem is pilferage. A few special rules, however, can keep pilferage to a minimum:

- All kits should be kept as close as possible to a supervisory area, such as a foreman's desk.
- One person should be responsible for dispensing first aid supplies to individual work areas. The reserve stock should be kept under lock and key at all times.
- Employees must be made to realize that unnecessary removal of first aid materials, and even slow replacement of used contents, might mean the difference between life and death.

OSHA regulations prescribe the training of persons in first aid when no infirmary, clinic, or hospital is in proximity to the workplace. Good industrial practice calls for trained first aid personnel in all work situations, even if elaborate facilities are nearby.

The provision of first aid treatment space with at least a cot and supplies is highly desirable because the work of first aid people is helped by privacy and at least a minimum of equipment.

The regulations leave it entirely to the physician to approve first aid supplies. Various types of kits of supplies are commercially available. The best of supplies require good storage, and a systematic method of replenishing supplies after use or deterioration.

The one type of equipment specifically required by this section of the OSHAct regulations is that designed for quick drenching or flushing of body and eyes. Such facilities must be provided within the work area for immediate emergency use if employees are exposed to injurious corrosive materials.

The regulations do not spell out the details of these facilities, but certain characteristics are well known from experience, especially the experience of the chemical industry.

Eyewash fountains and emergency shows should be so distributed that no exposed worker has to travel far to reach them.

Supplies of water must be adequate (deluge flushing is the rule for emergency showers) and protected against freezing.

Controls should be such that they can be turned on to maximum volume with minimum effort. Persons with acid in their eyes must not be expected to grope for controls, and someone who needs to have corrosives washed off his or her body must not be obliged to understand complicated directions.

Showers and fountains, like fire protection sprinklers, must be protected against ill-considered actions, which might leave them without water supply just when they are needed.

The regulation appears to require fountains and showers in any situation in which exposure to corrosive materials is possible. This includes a host of presently unprotected shops as well as chemical plants, which have long known and used such means of washing off harmful materials.

## OSHA MEDICAL TRAINING STANDARDS

The OSHAct regulations for medical facilities are stated in three brief paragraphs. Their brevity should not mislead any employer into considering them insignificant. The full text of the standards promulgated is as follows:

a. Employer shall insure the ready availability of medical personnel for advice and consultation on matters of plant health.
b. In the absence of an infirmary, clinic, or hospital in near proximity to the workplace which is used for the treatment of injured employees, a person or persons shall be adequately trained to render first aid. First aid supplies approved by the consulting physician shall be readily available.
c. Where the eyes or body of any person may be exposed to injurious corrosive materials, suitable facilities for quick drenching or flushing of the eyes and body shall be provided within the work area for immediate emergency use.[1]

The standard requires that each employer make use of a consulting physician. In a large company or in one whose operations produces acute health hazards, the medical staff might well include one or more physicians, specialists in industrial medicine or in an area related to a hazard peculiar to the industry. In the great majority of companies, the medical consultant will be employed on a part-time retainer basis.

The specific duties assigned to the industrial physician will vary from company to company, but the language of the OSHAct regulation makes is clear that it is not enough to have a physician available only to treat accident victims and employees suffering acute illnesses.

The physician must consult and advise "on matters of plant health" and must approve the first aid supplies provided. The first requirement implies that the doctor must know the plant, its materials, its operations, and the

[1]CFR 1910.151.

hazards involved. And it is hard to imagine how he or she could meet the requirement of approving first aid supplies without knowing the first aid facilities, qualifications of the first aid personnel, and the types of injury and other health problem that might be expected in the plant.

## INDUSTRIAL HYGIENE TRAINING

Medical and industrial hygiene facilities overlap to some extent, and they are certainly interrelated. Hearing conservation is one of the most obvious of the activities that are related to both medicine and hygiene.

Often it is the industrial nurse who heads the on-the-premises medical team; the physician is available as a consultant and makes examinations on a regular schedule.

OSHAct regulations make it imperative that there be interdisciplinary cooperation between the safety professionals, industrial hygienists, and medical personnel.

There are three key concepts in an effective industrial hygiene program—recognition, evaluation, and control.

- Recognition requires the knowledge of the stresses arising out of environment, operations, and processes.
- Evaluation involves a judgment or decision that is usually based on a combination of various measurements of the magnitude of the stress and on past experience.
- Control involves some or all of the following: isolating a hazard, substituting a less hazardous process, changing a process, training, housekeeping, the employment of exhaust and ventilation systems, and personal protective equipment.

The stresses creating hazardous conditions fall into one or more of the following categories:

- Chemical liquids, gases, dusts, fumes, mists, vapors—that is, respiratory, contact, or ingestion hazards
- Physical—electromagnetic or nonionizing and ionizing radiation, noise, vibration, temperature extremes, and pressure
- Biological—including insects, molds, fungi, bacteria, and viruses; of particular concern are viruses carried in blood or other body fluids such as hepatitis B virus and the acquired immune deficiency syndrome (AIDS) virus.
- Ergonomic—including matching mental and physical requirements of a job (e.g., operating a large machine such as a lathe) to human capabilities and staying power to minimize monotony, fatigue, and so on

Of the three modes of entry of a toxic substance—inhalation, ingestion, and absorption—probably the most important is inhalation, because of the rapidity with which a toxic material can be absorbed into the bloodstream from the lungs.

## A COOPERATIVE TRAINING EFFORT

No single group can solve all of the many training problems presented by occupational health hazards. Before any evaluation of an industrial health hazard can be made, there must be an understanding of the whole person. It is not enough to know where and when a hazard occurs on the job and the employee's location or movement in relation to the hazard; there must be some consideration of the employee's off-the-job activities. For example, it is not enough to know how many decibels an employee is exposed to on the job (and for how long each day); it is important to also know what kind of noise the employee is exposed to off the job.

It is difficult to measure off-the-job exposures to hazardous materials or situations, but these exposures cannot be ignored. Some clues come from biochemical measuring and monitoring. But don't overlook the information the employee can relate about him or herself.

Acceptable on-the-job exposures are spelled out for most hazardous substances in the form of threshold limit values (TLVs) published by the American Conference of Governmental Industrial Hygienists, 1014 Broadway, Cincinnati, Ohio 45202.

Where there is a potential for serious exposure, continuous monitoring is essential. Materials that do not represent too serious a problem, however, may be sampled periodically in accordance with the extent of the problem.

Recognition, evaluation, and control of hazards in an industrial hygiene program depend on cooperation with the other disciplines—engineering, medicine, and safety, as well as the purchasing department, line supervision, and the individual employees.

### Engineering

The introduction of new plant processes and operations usually begins with the engineering organization. For this reason, the engineering staff plays an important role in the control of occupational health hazards. It is part of their responsibility to plan all operations—using established engineering procedures—to prevent unnecessary exposure to harmful work environmental factors or stresses.

Engineers should notify the medical, industrial hygiene, and safety organization whenever new operations or processes are to be initiated. Equally important is a request for an industrial hygiene survey of new installations before permitting employees to operate new equipment or to work with new processes.

### Medicine

Medical personnel are responsibile for knowing the physical and emotional requirements of every job as well as the hazards involved with each job as factors in the placement of personnel whose physical and emotional capacities meet the minimum job requirements.

The medical team also has an obligation to develop adequate, effective measures to prevent exposure to harmful agents, to examine all employees periodically—especially those working with or exposed to hazardous materials, and to restrict employees from further exposure (on a medical basis) whenever warranted.

Often the industrial nurse acts as the liaison between the physician and representatives of other disciplines, and occasionally the nurse and the supervisor represent the employees to the other disciplines.

### Purchasing

The company's purchasing organization is responsible for making sure that the equipment and materials approved by industrial hygiene, medical, and safety personnel are purchased and that engineering specifications are included in the purchase order.

### Supervisors

Each supervisor is responsible for maintaining a safe working environment for employees in his or her charge. In addition to meticulous housekeeping, the supervisor must be aware of the hazards involved in each job, report any accidental exposures to hazardous materials, and know and abide by the work restrictions placed on personnel by the medical department. Supervisors must make sure employees have adequate training, have the proper personal protective equipment, know how to use it, and do, in fact, use it. Discipline and enforcement are usually delegated to the supervisor.

### Employees

Each employee is responsible for contributing to the overall success of a company's industrial hygiene program. Besides following work procedure rules, including the proper use of personal protective equipment, each employee has the obligation to report any accidental exposures or any hazardous conditions that may arise.

### Safety

Probably the safety organization plays the key role in the overall environmental health program—that of coordinator.

It is the safety team's basic responsibility to conduct an effective safety program, but this can be done only with the cooperation of engineering, medical industrial hygiene, purchasing, and supervisory personnel.

Safety's role is educational in that employees must be adequately trained with respect to the hazards involved in their work, regardless of the other requirements of the job. Rules, regulations, procedures, and enforcement are usually initiated by the safety personnel, but much of the responsibility entailed can be delegated to supervisory personnel.

Since new materials and new processes are constantly appearing, an industrial hygiene program remains an ongoing program. Fortunately many organizations as well as consultants specialize in this field. Insurance companies, manufacturers, private laboratories, and consultants can provide a company-wide survey to identify, monitor, and control work environment exposures to health hazards.

## MEDICAL PROGRAMS

The following is reproduced from the OSHA Safety and Health Program Assistance Series, Pamphlet 11 (September 1991 Draft), produced by the Office of Consultation Programs. This series of pamphlets represents some of the most useful information produced by OSHA.

**The Occupational Health Delivery System: A Medical Program That Is Right for Your Worksite**

### INTRODUCTION

Are you forgetting the **health** in your occupational safety and health program? The Occupational Safety and Health Act of 1970 has as its goal "to assure so far as possible every working man and woman in the Nation safe and healthful working conditions . . ." [29 U.S.C. 651, Sec. 2 (b)]. The OSHA Safety and Health Program Management Guidelines (54 FR 3908-3916, January 26, 1989) strongly urge the identification and control of health hazards and the implementation of a medical program.

A medical program is another name for the systems that employers put in place to ensure occupational health expertise is used in their overall safety and health programs. Having a medical program does not necessarily mean that you must go out and hire a doctor to work at your company. There are many ways for you to find and use occupational health expertise. This pamphlet is aimed at helping **you decide what will work best for your business.**

In this pamphlet we will call the medical program the "occupational health delivery system," or OHDS. This term will help you remember that a comprehensive program is more than a *response* to work-related injuries and illnesses *after* they happen. It also means those activities which you do to find the safety and health hazards in your business and your plan for their prevention or control. It is a management system in the same way that the activities you do to promote safety and health are a management system. After all, your safety and health system is much more than the office which is occupied by the person in your company who is responsible for safety and health. Similarly, your OHDS is more than the place in your business where you keep the first aid kit.

If you find it more difficult to establish the goals and objectives for your OHDS than for the other parts of your safety and health program, it may be because the harm it prevents may not, at first, seem obvious. For example, an employee who has hand pain and is gradually developing a cumulative trauma disorder (CTD) will seem to have a less serious problem than the employee who has a severe cut or a broken bone from an accident. But OSHA's experience is that work-related health problems are no less serious in terms of costs and human suffering than the more obvious injuries.

An effective OHDS will help reduce all types of safety and health hazards and the resulting injuries and illnesses. The positive results from such a program will be measurable by a decrease in lost work days and workers' compensation costs. They will also help increase worker productivity and morale.

## WHO SHOULD MANAGE THE OHDS?

You will find that having occupational health professionals (OHPs) manage the OHDS works best. This means a physician or a registered nurse with specialized training, experience and knowledge in occupational health works with you but *not necessarily as your employee.* This arrangement works best because safety professionals, industrial hygienists, occupational medicine physicians and occupational health nurses each have their own area of specialized knowledge. You cannot expect to get all the information and service you need for your safety and health program from only one type of specialist. If you tried, you might miss identifying a dangerous hazard in your business.

In Appendix 3 at the end of this pamphlet, you will find a description of the different ways that physicians and registered nurses receive specialty training in occupational medicine and health and the different services that they can provide for your business. In Pamphlet 8 in this series, "Establishing Complete Hazard Inventories," there is information about some of the services you can get from safety professionals and industrial hygienists.

## WHAT SERVICES DO YOU NEED FROM YOUR OHDS?

### What to Consider

There is no such thing as a standard OHDS. As an employer or manager for a business, you need to think of the special characteristics of your business when developing an OHDS that is right for you. These special characteristics include:

- The actual processes in which your employees are engaged
- The type of materials handled by your employees

- The type of facilities where your employees are working
- The number of employees at each site under consideration
- The characteristics of your work force, such as age, gender, ethnic group, and education level
- The location of each operation and its distance from health care facilities

As you look at the characteristics of your worksite, you should be asking yourself questions such as:

- Are there hazards in our processes, materials or facilities that make it likely that our employees will get sick or hurt?
- Do we have so few employees that on-site occupational health resources are less practical than off-site contract services? Do we have so many employees that we will save time and money by installing on-site resources?
- Are there special characteristics of our workers that would make them more vulnerable to illness or less likely to understand the dangers of illness?
- Is there anything about our workplace that makes it important to have occupational health assistance nearer to us or faster to reach?

When you examine your business with these questions in mind, you will be better able to choose which OHDS services you need. The services are listed below. Appendix 1 includes some examples of how different companies of varied sizes tailor OHDS activities in their businesses.

## THE FULL RANGE OF OHDS FUNCTIONS

There are three basic types of OHDS activities:

1. Prevention of hazards that cause illnesses and injuries.
2. Early recognition and treatment of work-related illness and injury.
3. Limiting the severity of work-related illness and injury.

### Preventing Hazards

- Make sure that your safety and health policy shows that you are as concerned about your employees' health as their safety. (See Pamphlet #2 of this series, "Safety and Health Policies.")
- Make sure that qualified OHPs help you identify the hazards and potential hazards of your workplace. (See Pamphlet #8 of this series.)
- Use OHPs in the development and presentation of health training. (See Pamphlet #12 of this series, "Safety and Health Training" and OSHA publication #2254 (reprint), "Training Requirements in OSHA Standards and Training Guidelines.")

- Provide professional occupational health expertise as a resource to your safety and health committee. (See Pamphlet 4 of this series, "Employee Involvement.")
- Be sure to include your OHDS in your annual self-evaluation to ensure the quality it provides. (See Pamphlet #7 of this series, "Determining How Well Your Safety And Health Program Works.")

### Early Recognition and Treatment

- Use OHPs to help you decide, on the basis of existing or potential hazards at your workplace, when baseline and periodic testing of your employees and new hires for exposure evidence may be necessary. This is called "health surveillance" and is required by some OSHA standards for specific types of exposures.
- Use OHPs to do the testing needed for health surveillance.
- Make sure records are kept of employee visits to first aid stations, nurse's office, contract clinic or hospital and have an OHP review the symptoms reported and the diagnoses to see if there appear to be patterns that indicate an occupational health problem.
- Provide first aid and CPR assistance through properly trained employees on every shift, making sure that employees keep up their certifications and that they are adequately trained in the specific hazards of the worksite. Also make sure that they understand the hazards from blood bearing communicable diseases, including hepatitis B and AIDS, and how to protect themselves. (See note in Appendix 3.)
- Make sure that the OHPs whom you use have current credentials, have had recent occupational health continuing education, and understand the hazards of your worksite, so that they will be able to recognize **early** symptoms of occupational health problems and begin prompt and appropriate treatment to prevent disability.
- Make sure that standardized procedures (called "protocols" in the medical community) are used in all aspects of occupational health delivery systems, particularly if you are using more than one contractor for health services. (For more information, see Appendix 5, "Protocols. . . .")
- Have one of the OHP's keep your employee injury and illness records. (See Appendix 2, "Recordkeeping.")

### Limiting Severity

- Coordinate the response to emergencies of all the responsible individuals or departments that are at your worksite and of all the emergency organizations off the worksite, such as the fire department, any contractual organization or a near-by community hospital, so that everyone knows exactly what to do and what to expect from others. (See Pamphlet #10 of this series, "Establishing Hazard Correction and Control Programs," and OSHA publication #3088, "How to Prepare for Workplace Emergencies.")

- Maintain contact, through your OHP (whether your employee or arranged by contract), with any employee who is off work due to an occupational illness or injury keeping in touch with the practitioner providing treatment and care to ensure that the treatment is appropriate and that the employee is responding as expected.
- Use your registered nurse or physician contact with an employee who is off work for an extended period to help advise the employee about Workers' Compensation benefits and rights and about ongoing care.
- Use these OHP's to provide evaluations aimed at determining whether an employee can resume full duty after injury or illness or whether their work duties need to be modified.
- Consult your physician or registered nurse for help with the development of the modified duty position, to ensure that the employee can perform the work and get the benefits of feeling productive again.

### SUMMARY

Your business can use an OHDS to deliver a number of important services. As an employer, you need to consider the special characteristics of your business in order to determine which of these services are appropriate. The factors involved in making this decision include the type of processes and materials your employees work with and the resulting or potential hazards. Other things to think about are the type of facilities in which they work, the number of employees at each site and the characteristics of this work force, such as age, gender, cultural background and educational level. The location of each operation and its nearness to a health care facility is also important.

## APPENDIX 1

## EXAMPLES OF OHDS USE

Below we have presented three examples of the organization of the OHDS at large, medium and small workplaces. The first is a manufacturing firm with about 1800 employees, divided between two locations. The second is a small, independent janitorial service with 35 employees. The last example is a meat packing company with about 500 employees. These examples show that each company's service needs are individual and that their methods for providing the services differ. The key issue demonstrated in the examples is that sometimes services are best provided by people that you hire as your own employees. In other situations, it may be better to contract for these services. In our examples, you will notice that sometimes employers used community agencies to supply services. Using community agencies may save you money.

### Background Information on the Sample Companies

#### *Example 1*

One thousand of the workers employed by the larger manufacturing company are at one site, where the components of their major product are manufactured. A second site, five miles away, has 500 employees and includes a line where the components are assembled. Also in that building is the facility's maintenance department and a garage where the company trucks and other vehicles are parked and serviced. At that same site, but in a different building, are the executive offices which house all the administrative divisions, such as accounting, human resources, marketing and a small show room.

The manufacturing site is 17 miles from the nearest medical facility. The light assembly and administrative site is 12 miles from the nearest medical facility. The company runs two, eight-hour shifts. These are a day shift and an evening shift. There are two security personnel working alone from midnight to 7:00 a.m.

Company One employs a safety professional to head its safety department. This company employs an occupational health nurse practitioner to head its OHDS. Both departments have other professional and non-professional staff to support the department heads. The safety department and OHDS both have offices at the manufacturing site.

#### *Example 2*

Company Two, the small, independent janitorial agency, provides cleaning and light maintenance services for commercial buildings. There are 35 employees, including the owner-manager. There are three clerical support personnel, an evening supervisor and 30 service personnel, 20 men and 10 women. The service personnel report to a central office from which they are dispatched in teams of two or three in company vans. The service personnel all work an evening shift from 5:00 pm. to 1:30 am. Two of the clerical employees work a day shift, 9:00 a.m. to 5:30 pm. One clerical employee works a shift that spans the day and evening shift. The supervisor works the same evening shift as the service personnel.

No safety, industrial hygiene or occupational health professionals are employed by this company.

#### *Example 3*

Company Three, the meat packing company which employs 500 workers, has 460 people working in the slaughter department and the department where the carcasses are dismantled. These divisions operate on a 10 hour per day shift. In addition, there are 20 people who perform plant sanitation and maintenance functions on an overlapping evening shift. There are also 20 employees functioning as supervisors and in staff positions such as personnel, payroll, and safety and health.

The plant is seven miles from the nearest health care facility. The company employs a full-time occupational health nurse (registered nurse) and a full-time safety director.

In the sections below you will find major portions of an OHDS and descriptions of the way they are handled by each of our three sample workplaces.

### Hazard Analysis

*At company one*, the basic work to provide a comprehensive assessment of hazards (see pamphlet 8 of this series) was done by a committee composed of the safety director, the occupational health nurse practitioner, the director of manufacturing, the supervisor of maintenance and two line employees, one from day shift and one from evening shift. In addition, a consultation was requested from the liability insurance company's loss control division. The consultation was conducted by an industrial hygienist who confirmed the need to monitor for noise in the manufacturing area. She also helped the occupational health nurse practitioner and the human resources director write job descriptions for the major employee classifications. The descriptions emphasized important safety and health considerations such as the amount of weight lifted and the chemicals handled.

A second consultation was obtained from the department of occupational medicine at the near-by university. The occupational medicine physician suggested a design for an optional health surveillance program for cumulative trauma disorder (CTD) in the shoulders, arms and hands of the employees in light assembly. The program is designed to use personnel and other resources already available at the company.

*At company two*, the owner-manager of the janitorial service was aware, from reading the newspaper, that OSHA was enforcing protective measures for workers who were exposed to ergonomic hazards. She contacted the OSHA-funded, state-run consultation service and received information about the criteria which were being used for enforcement. This information guided her in organizing a management system which would meet the enforcement requirements. At the owner's request, consultation personnel examined this system and other aspects of the company's safety and health program, including the hazards to which employees were exposed. They helped develop protections against the hazards found and recommended a cumulative trauma disorder (CTD) training program with an emphasis on back injuries. They also recommended a driver safety program.

*At company three*, the meat packing company's hazard assessment included basic safety and industrial hygiene evaluations as well as a complete ergonomic review following the *OSHA Ergonomic Program Management Guidelines For Meat Packing Plants.* Their assessment included a check for potential exposures to communicable diseases from the slaughtered animals. The safety and health committee developed a list of questions about safety and health conditions and potential hazards in the plant. To answer these questions the committee performed a series of walk-throughs and employee interviews. The committee consisted of the safety director, the occupational health nurse, a supervisor, and four hourly employees, one each from the slaughter, fabrication, sanitation and maintenance departments.

In addition, the safety director and the occupational health nurse analyzed the actual jobs being performed by line employees for potential ergonomic problems. The OSHA Ergonomic guidelines helped them identify those positions which involved the specific activities associated with the development of cumulative trauma disorders (CTD). Furthermore, they reviewed all of the material safety data sheets (MSDS) for all the chemicals used for cooling and sanitation at the plant.

### Employee Training

*At company one*, the safety director and the occupational health nurse practitioner designed training programs to be delivered during employee orientation to inform the employees about the company safety and health policies and program, and to alert them to the specific hazards in their jobs and what they need to do to protect themselves. The training is conducted by safety department and OHDS staff members who have taken special courses to prepare them as trainers. The employees are invited to contact the safety director, the occupational health nurse practitioner or their supervisors if they have further questions. The training includes a short test at the end to demonstrate that the employees understand their specific risks and how to protect themselves. There is a regular schedule for follow-up training.

Because the plants of company one are more than ten minutes away from the nearest medical facility, the decision was made to establish emergency response teams in both locations and on both shifts. The teams are set up on a volunteer basis and consist of five employees per shift per location and all the security personnel. The company contracts with the American Red Cross to provide the first aid and cardiopulmonary resuscitation (CPR) training and refresher classes at the company.

*At company two*, the state OSHA consultation office suggested contracting with a nearby occupational health clinic, already providing work-related employee health services, to conduct training in the prevention of CTD with an emphasis on back safety. In addition, the employees are informed about the contents of the cleaning solutions they are using along with proper mixing techniques and the use of gloves and protective eye glasses. The material safety data sheets (MSDS), which list toxic ingredients, are explained and employees are told where the MSDSs are kept (See Appendix 2, Recordkeeping, for a description of MSDS.)

The initial class was provided at the janitorial service company's central office. New employees are instructed at the clinic as part of their preplacement physical examination and orientation. Finally, the owner received booklets from the Automotive Occupant Restraint Council, which promotes safe defensive driving and the use of seat belts. These booklets were distributed to all current employees and are included in the orientation materials for all new employees.

The employer encourages the service personnel to take a beginning first aid course which is offered through a local municipal adult education program by granting paid-time for the class. The evening supervisor was required to take the beginning and advanced first aid course. Each van is supplied with a first aid kit, as is the office. The service employees are instructed to report all injuries and illnesses to the supervisor at the end of their shift, and sooner by phone if they think that the problem needs more than minor first aid treatment. The supervisor then refers the employee to a nearby emergency room or the contract occupational health clinic as necessary.

*Management in company three* decided to conduct the employee training program using their own company personnel. The safety director designed the training to address hazards such as fire, walking surfaces, cuts, elevations and ammonia leaks. The occupational health nurse designed training to address hygienic practices to reduce the possibility of exposure to the biologic hazards such as brucellosis, anthrax and Q fever associated with animal handling. In many respects, the activities performed by the workers to protect the meat from contamination, as required by the USDA, also protects them. Where this is not the case, the program is designed to emphasize what employees need to know and do to protect themselves. The nurse also developed material which informs the employees about the early signs and symptoms of cumulative trauma disorders (CTD) and ways to help prevent them.

Together, the occupational health nurse and the safety director at the meat packing company instruct the employees about the chemical and temperature hazards associated with the industry. Additional classes are held for the clerical, sanitation and maintenance workers. In addition, supervisors are trained about the early symptoms of ergonomic problems, so that they will encourage workers to report these problems as readily as other injuries or illnesses.

Five volunteers from each plant area or unit and all the supervisors from each shift were trained in first aid. Commitment to be on the emergency response team is for one year. Retraining occurs for all team members at the anniversary date, when new members are added.

### Health Surveillance

*In company one*, machines in two departments were extremely noisy. The safety director designed and the maintenance department constructed double layered sheet rock walls with sound reducing baffles between them around the two machines. Then the company contracted with an industrial hygienist to perform an environmental sound survey. This survey showed that the noise level in one of the departments was still too high, even after the construction of the sound baffles. The company also contracted with a near-by audiologist to perform baseline pure tone hearing tests on all current employees. New employees for the department designated as too noisy are tested as part of their orientation. The required annual audiometric testing is done by this same audiologist. The occupational health nurse practitioner conducts the education program about hearing conservation. She also did the research necessary in order to purchase the best hearing protectors for the employees.

*Company two* had no need of health surveillance.

*Company three* consulted with the occupational medicine physician to develop a surveillance program for cumulative trauma disorder (CTD). As a result the physician made the portion of the pre-placement physical examination that dealt with the upper extremities and the back more detailed for the workers in the slaughter department and the division where the carcasses are dismantled. At the six month and one year anniversary dates, randomly selected employees are invited back to be reexamined. Examination results which indicate early development of CTD are reported to the employees, and management is informed about which positions need further evaluation. No personal information which identifies particular employees is given to management, however.

### Analysis of Employee Use of the OHDS

*At company one*, a clerk in the OHDS is assigned to enter information about each employee visit to the OHDS onto a spread sheet which includes the date, the time of day, the employee's department, the employee's complaint and the treatment rendered to the employee. Totals from the graph are examined by the occupational health nurse practitioner every month. Any unusual clusters of complaints are investigated by the safety committee.

*For company two*, the contract occupational health clinic sends a monthly statement to the employer which summarizes all bills which have been submitted to this company's Workers' Compensation insurance carrier. This summary includes both diagnostic and treatment information. The employer analyzes this information for trends in injuries and illnesses as one way of deciding if employees are being exposed to identified hazards, if hazards exist that have not been identified or if employees needed more training about hazards.

*At company three*, the occupational health nurse maintains a spread sheet of all employee visits to the health office. She combines this information with that received from the contract occupational medicine physician to form a report which is presented to the safety and health committee each month. This report and the accident reports become the basis for special safety and health emphasis programs within the company.

### Establishing Standard Procedures

*For company one*, the occupational health nurse wrote procedures for the "first aiders" to use when administering first aid, CPR and emergency transfer of ill or injured workers. She also wrote procedures which described the standardized assessment and onsite treatment that she uses for employee illnesses or injuries and for all health surveillance programs.

*For company two*, the state OSHA consultation staff worked with the owner to develop standardized procedures for first aid and emergency situations. They also wrote specific procedures for mixing and use of all the cleaning solutions, using the buddy system for lifting heavy objects and a system for task rotation for lengthy repetitive activities such a vacuuming. Discussion of the procedures is included in the employees' orientation.

*At company three*, the occupational health nurse and the contract occupational medicine physician worked together to write procedures for all the treatment given in the OHDS office, including the dispensing of over-the-counter medications. Procedures were also written for hygienic practices for the employees exposed to biologic hazards which ensured that they did not consume food, beverages and tobacco products with contaminated hands, and that they did not accidently contaminate their street clothes or shoes before leaving the plant. The safety director and the occupational health nurse developed procedures which ensured that the employees of contractors performing pest eradication operations did not accidently expose employees to pesticides. Finally, this same team developed procedures which included proper work techniques and frequent knife sharpening to prevent ergonomic problems.

### Early Recognition and Treatment

*At company one*, the occupational health nurse practitioner who heads this department has graduate level training in assessment and management of occupational illnesses and injuries. She is licensed to treat many of the employees' occupational injuries and illnesses using previously approved standard procedures. A referral relationship has been established by contract with a local hospital that has an emergency room and an occupational medicine clinic. Employees with illnesses or injuries that are assessed by the nurse practitioner to be too severe to be treated onsite are transferred to the emergency room. Those employees who are receiving treatment by the nurse practitioner and do not respond as expected are referred to the occupational medicine clinic. In this way, a majority of the company's work-related injuries and illnesses are treated within the OHDS.

*At company two*, the contract clinic's medical director has completed a mini-residency in occupational medicine and has ample knowledge of the risks to which this company's workers could be exposed.

*Company three* has a policy that encourages employees to promptly report symptoms of illness and injuries to the occupational health nurse. The nurse treats minor illnesses and injuries in the plant health office using simple dressings, ice and over-the-counter medications. She refers more severe problems to the contract physician. This nurse has taken a continuing education course in the recognition and conservative treatment of CTDs and is able to implement treatment and referral early. She is also able to review preventative measures with the employees at each visit.

### Medical Case Management

*At company one*, the occupational health nurse practitioner works with the personnel manager to develop a case management system for all employees who are off work with illnesses or injuries lasting more than five days. The system consists of a method for prompt treatment authorization, a referral list for second opinions, assistance in filling out insurance forms, communication with the insurance carrier to ensure timely benefit payments and ongoing contact with the employee and the family to answer questions.

*At company two*, the owner functions as the human resources director as well as the manager. As such, she is in close contact with any employee who experienced lost work time related to industrial injury or illness. She considers this adequate case management. Her workers' compensation insurance carrier assisted her by providing information about helping injured and ill workers return to work quickly. The suggestions prompted her to increase the frequency with which she makes telephone contact with those persons.

*At company three*, the occupational health nurse practitioner and the contract occupational medicine physician maintain close communication about all employees with work-related injuries and illnesses that are not responding to treatment as expected. The occupational health nurse practitioner ensures that specialist referrals occur promptly. The practitioner also works closely with the supervisors so that employees can return to modified duty positions as soon as possible by clearing the proposed positions with the treating physician.

### Coordination For Emergency Services

*At company one*, the safety director, the occupational health nurse practitioner and the head of the security department worked together to develop a system whereby all employees in each department know their exact responsibilities in the event of an emergency. They also discussed their plan with the local fire department that will be responding to emergency calls.

*At company two*, because of the nature of the operation, with employees moving from one workplace to another instead of having a fixed worksite, no special arrangements were made with emergency organizations. The company, however, did make its own emergency preparations (discussed under "Training.")

*At company three*, the occupational health nurse practitioner and the safety director head the team and respond to each emergency. The plant receptionist is responsible for contacting outside emergency organizations, so he is included in the emergency response team meetings.

### Recordkeeping

*At company one*, the OSHA 200 Log, the material safety data sheets and the results from the noise surveillance are maintained and kept in the OHDS office where the occupational health nurse practitioner can answer employee questions. All employee visits to the OHDS are documented in the individual employee medical record.

*At company two*, the OSHA 200 Log and the material safety data sheets are maintained by the owner and are available for the employees to see upon request.

*At company three*, the OSHA 200 Log, the material safety data sheets and the results from the CTD surveillance are kept in the OHDS office where the occupational health nurse can answer employee questions. All employee visits to the OHDS are documented in the individual employee medical record.

(More detailed information about recordkeeping is presented in Appendix 2.)

## SUMMARY

The examples demonstrate how three different employers provide OHDS services using a combination of resources within their companies, subcontractors and governmental agencies. Some of the sub-contractors brought the services to the premises, while in other situations the employees traveled to the contractor to receive the service. In each case, the services provided were selected based on the special characteristics of the business process, the potential exposures within that process, the business location and the employee population.

The OHDS in the examples also demonstrate the use of activities aimed at the prevention of exposures, the early recognition and treatment of work related illnesses and injuries and those activities that decrease the severity of and potential for disability from work related illness and injury. The inclusion of all three types of activities produce the most safe and healthful work environment.

## APPENDIX 2

## RECORDKEEPING

OSHA requires that you keep certain records. Employee visits to the health office will generate other records. Maintaining your records using the resources of your OHDS is a good option because the occupational health professionals connected with your OHDS have the training to answer your employee's questions about these records and decide whether their complaints are work-related. Please remember, in addition, that individual employee health records contain confidential information which may be protected under case law relating to the right of privacy. Access to these records should be managed by health care professionals.

What follows is a description of the health-related records that you should have in your business. It is meant to give you enough information to decide if you want to manage these records within your OHDS. You need to consult the standards or the other references for more details. The Code of Federal Regulations (CFR) numbers or publication titles have been included for your convenience.

*"OSHA-200 Log"*—OSHA requires that you keep track of all the work-related injuries and illnesses that occur at your work sites. OSHA Form 200, often called the OSHA-200 Log, is used for that purpose. Another form, The Supplementary Record of Occupational Injuries and Illnesses (OSHA form number 101), or its equivalent, must also be kept. Employers often use the "First Report of Injury," required by the Workers' Compensation system, as this supplementary record.

The OSHA Recordkeeping Standard, 29 CFR 1904, also tells you how long you must keep the records, and how you are to inform your employees about the injuries and illnesses occurring in your company. For more information, see this standard and the Bureau of Labor Statistics 1986 guidelines entitled, "A Brief Guide to Recordkeeping Requirements for Occupational Injuries and Illness" or the more comprehensive "Recordkeeping Guidelines for Occupational Injuries and Illnesses" (both are OMB #1220-0029).

*Environmental and Employee Health Monitoring Records*—When you do any type of environmental monitoring, such as air sampling, OSHA requires that you keep the results. If you test employees for exposure to hazards, you must keep this information also. The publication, "Access to Employee Exposure and Medical Records," 29 CFR 1910.20, tells you that employees and their representatives must be allowed to see and copy the records, and tells you how long the records must be kept.

*Records Required for Specific Hazards*—OSHA has issued a number of standards that tell you things that you have to do when your business involves certain hazards. Often this includes testing the work place and employees for signs of the hazard. When this is the case, the standard also tells you how long you have to keep the results of the tests. Two examples of this type of standard are the asbestos standard, 29 CFR 1910.1001(m), and the occupational noise exposure standard, 29 CFR 1910.95.

*Hazard Communication*—The Hazard Communication Standard addresses more than recordkeeping. It tells you how you must communicate to your employees and your community about the chemicals you use or make in your business. But, because the standard requires that you keep certain documents and communicate certain information, it is discussed here.

The standard says that you must, among other things, keep a list of all hazardous chemicals present at your work sites. You must have labels and signs to warn people about these chemicals. You must have material safety data sheets (MSDS) for all hazardous chemicals. MSDS's provide information from the manufacturer about what is in a chemical and its health effects. In addition, employees must be trained about the chemicals to which they are exposed. The training must include information about what you are doing to protect employees and what they need to do to protect themselves. (See 29 CFR 1910.1200.)

*Individual Employee Health Records*—Employee visits to the health office will generate records which should be kept in order to document what treatment is being provided and how the employee responded. These records help improve the consistency of the health care. They can also be analyzed to discover possible unrecognized hazards, as described earlier in the section about analysis of employee OHDS visits.

# APPENDIX 3
# QUALIFICATIONS OF OHDS PERSONNEL

Once you have decided which safety and health services you want to provide through your OHDS, you need to decide who will provide the services. There are several factors to consider in making this decision.

First, your OHDS must be organized so that the personnel providing the services are not working alone when state law requires that they be supervised by a registered nurse or physician. Second, the occupational health professionals you use must have specialized, up-to-date training or experience in the methods of occupational health care. Finally, you must choose whether to hire OHDS personnel as your own employees or whether to contract for their services outside your company.

## Occupational Health Professionals

Occupational health professionals are medical doctors (MD), doctors of osteopathy (DO) and registered nurses (RN). They hold a license to practice their professions and they practice independently under this license, using the type of standard procedures described in this pamphlet.

*Occupational Medicine Specialists:* Occupational medicine specialists are medical doctors (MD) or doctors of osteopathy (DO) who have additional training or experience in the treatment of work-related illnesses and injuries. This training or experience can be acquired in different ways. A number of universities offer residency programs which will enable MDs and DOs to become board certified in occupational medicine through the American Board of Preventative Medicine. Professionals graduating from medical school or osteopathic training before January 1, 1984 may qualify for board certification by the "alternative pathway" which requires a combination of education and experience practicing in occupational medicine.

There are shorter programs, called mini-residencies, which offer academic training in occupational medicine. In addition, MDs and DOs can obtain advanced education in occupational medicine by taking continuing education courses. There are MDs and DOs working in occupational medicine who have gained advanced knowledge through experience working in the field for extended periods. A qualified occupational medicine specialist should be capable of performing or managing all the OHDS activities described in the previous section.

*Occupational Health Nurses and Nurse Practitioners:* Occupational health nurses are registered nurses (RN) who have received specialized training in occupational health. Like occupational medicine physicians, occupational health nurses may gain this training in formal college programs, through continuing education or through experience working in the field. Professional registered nursing licensure (RN) allows nurses to perform many health evaluation and care functions independently. An occupational health nurse should be capable of performing or managing most of the OHDS activities described in the previous section.

Nurse practitioners are registered nurses who have completed formal advanced training in physical assessment and the management of minor, stable illnesses and injuries. In some states, they are certificated for advanced practice by state licensing boards.

Nurse practitioners perform many health evaluation and care activities independently. They may perform many other such activities, under the direction of a physician, through the use of the type of standard procedures known as protocols in the medical community. The physician is not required to be physically present. Nurse practitioners can also take additional training and specialize in the treatment of occupational illnesses and injuries. When working with standard procedures, nurse practitioners should be capable of performing all the OHDS program activities described above.

### Support Personnel

Support personnel can provide more limited services. They have received specific training and are usually certificated or licensed by the educational institution where they received the training. Sometimes licenses or certification are granted by the state. The scope of practice for support personnel requires that they work under the supervision of licensed health professionals except when delivering first aid. Licensed vocational nurses (LVN), licensed practical nurses (LPN), emergency response personnel and first aid personnel are in this category.

*Licensed Vocational or Licensed Practical Nurses:* Licensed practical nurses (LPN) or licensed vocational nurses (LVN) are licensed by state agencies to perform certain health care activities such as taking blood pressures and applying dressings. They must practice under the supervision of a physician or a registered nurse.

*First Aid and Emergency Response Personnel:* A person does not need a formal health care educational background to be trained in first aid and cardiopulmonary resuscitation (CPR). A certification in first aid is usually granted by the training provider offering the training, after the student completes a standard curriculum and demonstrates competence. The American Red Cross is one example of an institution offering this type of instruction. "First aiders" perform treatments such as splinting and application of ice or pressure dressings. They also learn how to transport ill or injured people. "First aiders" fulfill the role of first responders. They can provide temporary treatment until care of the ill or injured person is transferred to someone with more advanced health care training. When employers use their employees as "first aiders," the employees should be properly trained and certificated because even first aid techniques can be harmful when done incorrectly. The training should be updated on a regular basis.

The teaching of cardiopulmonary resuscitation (CPR) is sometimes included in first aid courses but may have to be taken separately. CPR is the act of providing temporary life sustaining artificial circulation and respiration when these functions have stopped, before and while the person is transferred to a medical facility. As with first aid, persons completing CPR training are certificated by the organization offering the training. CPR performed incorrectly can injure the person receiving it. When employers use employees as "first aiders" and expect them to perform CPR, the employees must be thoroughly trained, and this training must be updated at least annually.

**Please note: It is possible for persons rendering first aid or cardiopulmonary resuscitation (CPR) to be exposed to the viruses that cause Hepatitis B and Acquired Immune Deficiency Syndrome (AIDS). This is because the "first aider" can come in contact with body fluids, such as blood, which might contain these viruses. It is crucial that all first aid**

**courses, given to employees to qualify them to participate on company emergency teams, include information about how the employees can protect themselves from potentially infectious body fluids while performing first aid or CPR.**

Emergency response technicians or EMTs have received advanced training in first aid, CPR and other life support techniques. Within certain restrictions they can perform sophisticated emergency procedures and transport ill and/or injured people.

Records of the original training, licenses, update courses and certification of all employees participating in the delivery of occupational health services, including first aid, CPR and/or emergency response activities, should be retained within the OHDS.

### Summary

There are a number of types of health care providers that are qualified to perform services within your OHDS.

When considering the selection of an occupational health professional, it is important for you to make sure that their licenses are current and that they have received recent advanced training in occupational health care either through a residency or education program or through experience with enough time in the field to have advanced knowledge.

When considering how to use support personnel, it is important for you to make sure that their licenses or certificates are up-to-date. It is also important to make sure that they are practicing within their scope of practice and under proper supervision. (See Appendix 4 for information to use in evaluating health care personnel.)

## APPENDIX 4

## EVALUATING THE QUALIFICATIONS OF A HEALTH CARE PROFESSIONAL

Whether you choose to employ professional health care personnel to develop and/or manage your OHDS or contract with outside vendors, it is important to evaluate their qualifications. Remember that occupational health professionals are people selling a service. You should use the same smart purchaser skills with physicians and registered nurses as you use with anyone who is trying to sell you something. This means knowing something about the services that you want to purchase. You can get this knowledge using this pamphlet and the other pamphlets in this series.

Here are some questions to ask prospective professional health care providers.

1. What type of training has the health care provider had?

   **Physicians:**

   Graduation date—note all degrees

   Request a copy of current MD or DO license

   Types of specialization certification

   Years of experience in occupational medicine

   Titles of continuing education courses taken in the last two years

**Registered Nurses:**
Graduation date—type of degree
Graduate degrees
Types of specialization certification
Years of experience in occupational health
Titles of continuing education courses taken in the last two years
Copy of current RN license

2. In what type of industries has the prospective health care provider had experience?
3. What kind of information does the prospective health care provider want to know about your business? A prospective health care provider should ask questions about the following:
   - your work processes
   - your known or potential hazards
   - your facilities, type and location
   - the number of employees
   - standards and/or regulations that apply in your business
   - health surveillance programs that have been done previously or are in progress
   - collective bargaining contracts
   - any OSHA citations previously issued
   - existence of a safety and health policy and its content
   - current method of providing OHDS services
   - other health care providers providing services
4. What does this provider know about OSHA recordkeeping requirements?
5. Given the above information, what would this provider do to contribute to the improvement of your safety and health program?
6. Can this provider provide references?
   - Has there ever been an OSHA inspection in a facility with which this provider was associated?
   - What was the outcome of that inspection?

## APPENDIX 5
## PROTOCOLS (ESTABLISHED STANDARDIZED PROCEDURES)

Protocols are documents which describe your standardized procedures, how you want something done in your company. You may already have standardized procedures for things like your accounting system or for servicing your equipment. Standardized procedures should be written for the activities of your OHDS. In your OHDS, they produce consistency in providing care for work related illness and injury, for reacting to emergency situations and for collecting data from health surveillance programs and all the other activities of your OHDS. Standardized procedures are not meant to interfere with an occupational health professional's treatment of work related injury and illness. They are aimed at ensuring the early detection of work related health problems through consistent and thorough evaluation of employee health complaints.

Standardized procedures also promote the treatment of work related illness and injury using the most up-to-date therapies. They are particularly important if you are using several contractors to provide your company's OHDS services, because they will help ensure that all your employees are receiving the same type of care. (See the Section of this pamphlet entitled "Examples of OHDS Use" for ideas of how and when employers sometimes use contractors.) Even if you have company employees providing OHDS services, there should be standardized procedures written for all health surveillance programs, health care and first aid given through your OHDS. These standardized procedures should be communicated to all health care employees and sub-contractors providing treatment for your workers.

The American Association of Occupational Health Nurses (AAOHN) has published *A Comprehensive Guide for Establishing an Occupational Health Service* which includes information on developing protocols. To obtain a copy of this guide, contact AAOHN, 50 Lenox Points, Atlanta, Georgia 30324.

## SUMMARY

Whatever the nature of the medical program and health delivery system, it is important that adequate emergency treatment be available and that medical and health services be available.

This usually requires first aid and CPR training for at least part of the workforce and coordination with in-house or nearby medical facilities.

This is an element of the safety and health program that cannot be neglected.

# CHAPTER 12

# FIRE SAFETY TRAINING

## INTRODUCTION

Fire continues to be one of the most dangerous hazards in the workplace, both in terms of personnel injuries and fatalities and property damage.

Safety training must include extensive information on how to prevent fires and the steps to be taken in case of a fire.

This chapter provides information that should be included in fire safety training.

## CAUSES OF INDUSTRIAL FIRES

Although many industries have specific problems of fire prevention and control because of the type of work performed or the materials used, the most frequent causes of industrial fires are common to all industries. Most of these causes can be controlled or eliminated. The success or failure of any plant safety and fire presentation program depends on the first-line supervisors. Here is a handy checklist of some things that can be done to help prevent a serious fire, which could destroy an entire plant; the percentages identify the extend to which a specific cause has contributed to industrial fires.

**Electrical Causes—22%.** Guard against electrical defects, mainly the kind that develop from lax maintenance in wiring, motors, switches, lamps, and heating elements.

**Matches and Smoking—18%.** Do not relax the rules in "no smoking" areas just because a fire has not happened yet. Be especially careful where flammable liquids are present, near stored combustibles, or in working areas where combustibles are used. If a department has a designated smoking area, be sure to restrict all smokers to that area; there can be no exceptions.

**Friction—11%.** Watch for hot bearings, misaligned or broken machine parts, choking or jamming material, or adjustment of moving parts.

**Hot Surfaces—9%.** Guard against exposure of combustibles to furnaces, hot ducts or flues, electric lamps or heating elements, and hot metal inprocess.

**Overhead Materials—7%.** Watch out for abnormal process temperatures, overheating of materials in driers, and overheating of flammable liquids.

**Open Flames—6%.** Be especially careful with gasoline or other torches. Stay alert for trouble with gas or oil burners.

**Foreign Substances—5%.** Caution operators to watch for foreign material in stock. Tramp metal produces sparks when struck by rapidly revolving tools. Oil pans used near cutting machines are a constant hazard.

**Spontaneous Heating—4%.** Watch for deposits in ducts and flues, low-grade storage, scrap waste, oil waste, and rubbish.

**Cutting and Welding—4%.** Supervise with care; these operations are highly dangerous in areas where sparks can ignite combustibles.

**Combustion Sparks—4%.** Instruct workers to be careful when burning rubbish; watch foundry cupolas, furnaces, and fireboxes.

**Miscellaneous—10%.** This includes incendiary cases (3%), fires spreading from adjoining buildings (2.5%), molten metal or glass (2%), static electricity near flammable liquids, as at spreading or coating rolls or where liquid flows from pipes (1.5%), chemical action (0.5%), and lightning (0.5%).

## FIRE PREVENTION PROGRAM COMPONENTS

The following factors should be considered in implementing a fire prevention program:

*Gas.* Gases of all types (oxygen, acetylene, methane, propane, butane, etc.) represent a hazard. All equipment (piping, regulators, and ignitors) should receive periodic and careful inspection by a qualified person. Employees must receive thorough instruction before using these types of equipment. Matches should not be used for lighting gas-fired equipment.

*Housekeeping.* Rubbish, waste, and other debris must be cleaned up and removed daily and disposed of in suitable containers outside of the plant.

*Electrical Equipment.* Inspection and maintenance of motors and other electrical devices should be done by a qualified person. All electrical equipment should be included in a periodic inspection.

*Matches and Smoking.* Smoking and the use of matches should be permitted in designated areas only. Oily or paint-soaked rags or clothing left in lockers are a great hazard. Dust, lint, and oil collecting in ducts and flues can also ignite spontaneously. Safety containers should be provided for rags and waste, and they should be emptied daily. Employees must be instructed carefully and supervised closely to prevent fires of these types. Ashtrays for personal smoking materials (e.g., matches, cigarettes and so on) should be conspicuously provided for employee use. Appropriate instruction in the use of these receptacles should reduce the likelihood of an accidental fire.

*Open Flames.* Open flames are involved in welding, forging, forming, and other heating operations. Fireproof materials around the work area and suitable clothing will aid in the prevention of most fires. However, the use of portable equipment involves high fire danger and all precaution should be taken, including having an extinguisher of the proper type immediately available.

*Heated Surfaces.* Heated surfaces on furnaces, flues, heating devices, and light bulbs can be the cause of fires if flammable or combustible materials are close enough to absorb sufficient heat to cause combustion. Care should be taken to ensure that all such devices are properly installed, especially with respect to clearance and barrier materials.

*Molten Metal.* Molten metal will ignite any flammable material with which it comes in contact. Precautions should be taken to provide a flameproof environment in foundry areas and also to provide flameproof clothing.

*Volatile Liquids.* Paints, varnishes, lacquers, petroleum products, solvents, and similar volatile materials are frequent sources of fire and explosion. The liquids themselves do not burn, but heat causes vaporization and the vapors of these liquids are easily ignited by sparks from electrical devices, static electricity, or metal objects striking together or, of course, open flames. Vapors are generally heavier than air and may travel along the floor for considerable distances before they are ignited and flash back to the source. Tight metal containers, flameproof cabinets, color coding, and strict storage and handling procedures are necessary. All persons in the plant must know exactly what to do in the case of fire.

## Smokers

Despite all cautions and warnings to the contrary, some persons persist in smoking, even in specially designated areas in which smoking materials present a clear hazard.

It is sometimes necessary to post signs indicating that smoking is or is not permitted. These signs should be discussed in training sessions and all employees made aware of the importance of self-imposed smoking discipline.

Smokers should be cognizant of various location and usage restrictions that apply throughout the company's facilities.

Smokers who persist in violating company rules relative to smoking must be dismissed or transferred in an effort to protect the individuals, their fellow employees, and company property.

### Emergency Response in the Event of Fire

Every employee must be given instruction in the following procedures:

- How and when to turn in a fire alarm
- How to extinguish fires and how to determine whether extinguishment should be attempted
- What do in the event of alarm (turn off all machines, disconnect power at main switch, turn off gas valves and torches, etc.)
- How to leave the building

A major fire is a disaster in any industry. Constant vigilance and quick action by the line supervisors can frequently contain a small blaze until the plant fire brigade or community fire department can extinguish it.

All supervisors are responsible for keeping areas around the fire doors and emergency exits clear at all times. Many plants have fire doors that close automatically to seal off an area to prevent the spreading of a fire. If these doors are blocked, however, they cannot function.

Another way supervisors can help to prevent a major fire is to make certain all employees know the location of all extinguishers and how to use them. This should be part of every indoctrination program for all new employees as well as the subject of regular review. Also be sure that everyone knows what type of extinguisher to use for a particular type of fire. Here is a rundown of fires and extinguishers of various types.

## FIRES AND EXTINGUISHERS

Traditionally, fires have been classified by their sources or fuels. Depending on whether the source that fuels the fire is an ordinary combustible, a flammable liquid, electrical equipment, or certain elemental metals, a particular portable fire extinguisher is recommended by the National Fire Protection Association (NFPA). Figure 12-1 presents the NFPA-approved classification system currently in use. This widely (if not universally) used system, recommended by the NFPA, is used to aid in ensuring the correct type of extinguisher is provided an dused based on the type of combustible materials present.

There are four major types of portable extinguishers available for purchase and use. Depending on the intended application of a fire extinguisher, it is necessary to be aware of the scope of products available for use.

No matter how extensive the testing of any mechanical device, there are still some hazards associated with their use, and operators of the devices should be advised of these hazards. Any training program related to the selection and usage of fire extinguishers should include material that describes their inherent dangers.

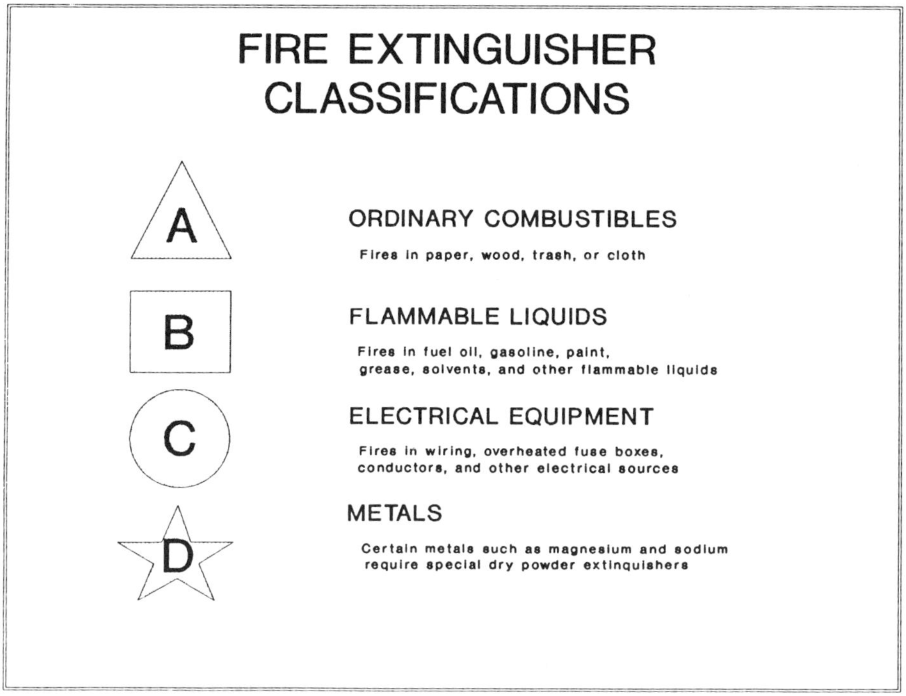


**FIGURE 12.1** Fire extinguisher classification system.

Operation of a portable fire extinguisher is a relatively simple task if one has been properly and completely trained. The correct method to be used varies depending on the class of fire (A, B, C, or D) and the type of fire extinguisher (water, dry, chemical-stored pressure, dry chemical-cartridge, or carbon dioxide).

### Extinguisher Maintenance and Instruction

Take a careful inventory of the extinguishers in each area and be sure that they are maintained properly. Many supervisors supplement the company fire brigade by training their own workers and drilling them to keep them sharp and aware of their assignments. By doing this, and by assigning men and women to particular extinguishers, a company can be certain that the right type of extinguisher will be available and that the person who is responsible for it knows how to operate it. One of the most popular workplace and home extinguishers is the all-purpose extinguisher that is effective for Class A, B, and C fires. However, there continues to be a need for instruction in recognizing the various types of fire and correct extinguishers for each type fire.

**Note:** Each state regulates its own rules for use of equipment. Be sure to check your own state requirements.

### Location and Identification of Equipment

Most extinguishers should be hung or placed so that their tops are not more than 5 feet from the floor. This permits ready removal without injury or accidental discharge.

Heavy extinguishers such as carbon dioxide units of more than 10 pound capacity, water or foam units of more than 2½ gallon capacity, and so on, should have their tops more than 3½ feet above the floor.

Extinguisher locations should be marked conspicuously. Signs or arrows should be higher than machines or stock piles for wide visibility.

Do not locate extinguishers in stair wells. Flue actions fans fire and may prevent access to the equipment.

Locate extinguishers close to likely hazards, but not too close. Accessibility and visibility are important considerations.

Avoid placement at locations where extinguishers are hard to reach, or where stock might be piled around them to block access or screen them from view.

Do not place equipment so that it extends out into busy aisles where people, trucks, or dollies might knock it down or damage it.

All extinguishers should be checked once a month to make sure the nozzles are not clogged, that they have been used but not recharged, or that they have not been damaged or tampered with.

All types of extinguisher should be checked and serviced semiannually or annually as directed on the label. All extinguishers should be recharged immediately after using.

Each extinguisher should be tagged to provide a record of its last recharge or inspection date.

## AUTOMATIC DETECTION AND SPRINKLER SYSTEMS

Most modern facilities are protected by automatic detection and/or sprinkler systems.

It is imperative that key personnel understand the capabilities, limitations, and basic methods of operating automated systems. This will generally require site and system specific instruction.

**Note:** Instruction is generally provided for new facilities or when new systems are installed. Unfortunately, in many instances where there is personnel turnover or the system has been installed for several years, reinstruction is not provided and building occupants may not know how to adequately check, test, maintain, or reset the system. They may also not understand the capabilities and limitations of the system and how to properly respond to alarms.

The most effective fire fighting system for many types of fires are sprinklers. Like automated detection systems, it is important that building occupants, especially key personnel, be knowledgeable of the system. Again, training should include the capabilities and limitations of the system and key operating and maintenance requirements.

For example, key personnel (or all personnel) should know the location of shutoff valves (both to check the status of the system and to shut off water flow in case of malfunction or damage to the system) and fire department connections. These are connections that allow local fire departments to hook pumper trucks and/or alternate water sources into the sprinkler system to increase water pressure and system effectiveness. They are usually on the exterior of the building.

All building occupants should also know not to store materials, move partitions, make building modifications, or otherwise block or mask sprinkler heads.

Supervisors and managers should be trained to understand the need to ensure that systems are properly designed, installed, and maintained. Fire protection engineers should be consulted any time the fire loading, configuration, or purpose of an area is changed.

## GUARDING AGAINST ELECTRICAL FIRES

The biggest single cause of industrial fire is electricity. Most of these fires stem from improper installation of electrical equipment and poor electrical maintenance. Annual property loss of $1.25 billion is due to electrical fires alone, and one factor that contributes to this loss is our tendency to forget that electricity, such a familiar and powerful production ally, is also a destroyer of production facilities. In addition, electricity is largely invisible; wires are generally concealed, and we take it for granted.

### Even a Light Bulb Is a Hazard

Have you ever seen someone burn his or her fingers changing a hot bulb? The amount of heat generated is pretty close to that of a match, and when a light bulb gets too close to combustible material, a fire can suddenly flash. That is why trouble lamps should have guards to prevent the hot bulb from accidentally brushing against a combustible. Open bulbs in a dusty atmosphere are another hazard. If dust accumulates on a bulb, the dust can char and drop as a spark with the bulb heats up. Prevent a fire by encasing all bulbs in dusty atmosphere in dust tight enclosures.

### Provide Proper Protection

Many times the installation of new equipment inadvertently produces an electrical hazard. Too often, this equipment is powered from old wiring that cannot carry the increased power load without overheating. This eventually leads to a short circuit and a destructive fire.

Never install, or permit others to install, an electrical circuit or wiring system without proper automatic overload protection. Most commonly used devices include fuses, circuit breakers, protective relays, and motor overload relays. They act automatically to deenergize a faulty circuit much as a safety valve depressurizes an overpressured tank. Never overfuse a circuit because there are frequent interruptions caused by a blown fuse or tripped breaker. These are danger signals and should not be ignored. Find the reason for the overload rather than trying to circumvent it. An overfused circuit eventually leads to a fire. One of the most effective overcurrent protection devices is the ground fault circuit interrupter (GFCI). These devices are highly recommended because they trip at a very low current flow (usually around 5 milliamps) providing not only excellent fire protection but personnel protection from electrical shock as well.

Take these five steps to provide proper maintenance for electrical protective devices:

- Keep them clean: remove dust, dirt, water, metal filings, or any foreign material.
- Tighten bolted, screwed, welded, or soldered connections.
- Lubricate to prevent frozen mechanical joints.
- Make sure ratings or settings on fuses, circuit breakers, relays, or other protective devices meet the design specifications of the circuits.
- Test protective devices under simulated fault conditions and compare their operations to manufacturer's specifications.

## HOW TO HANDLE HAZARDOUS GASES

Almost every production department in almost every industry faces the delicate materials handling problem of inflammable or explosive gases. Here are some tips on the proper storage, handling, and control of hazardous gases:

Dropping cylinders or allowing them to strike against each other may cause a mishap. Hold cylinders in place with chains or other suitable method to prevent toppling.

Oxidizing gases such as oxygen, nitrous oxide, and chlorine must never be stored with the easily oxidized gases such as acetylene, hydrogen, methyl chloride, ammonia, or liquified petroleum gas.

If cylinders are manifolded, provide adequate hydraulic flash arresters to prevent a flashback through the entire manifold system.

Keep outlet valves on cylinders tightly closed, even when considered empty. Use only tools and wrenches provided by the supplier of the cylinder valves.

Keep electrical equipment in storage areas at a minimum and use only equipment approved by Underwriters Laboratories (UL) for the type of gas stored.

A flashpoint is the lowest temperature at which vapors located above a volatile, combustible substance will ignite in air when the vapor–air mixture is exposed to flame.

Every substances has its own flashpoint, and persons working with such materials should receive training regarding appropriate temperature controls and flame avoidance.

## SUMMARY

Fire safety training is an important component of any safety and health training program.

Training for all employees should be site specific and should emphasize fire prevention and the steps to be taken in case of fire.

Training for supervisors and managers and other key personnel should also include training on automated detection and fire fighting systems.

Even though automated systems can greatly enhance the overall fire safety of an operation or facility, it is important that key personnel recognize the capabilities and limitations of the systems as well as operating and maintenance requirements.

## CHAPTER 13

# EVALUATING SAFETY TRAINING EFFECTIVENESS

### INTRODUCTION

This chapter presents a broad variety of methods of evaluating overall safety program effectiveness as well as the components of a safety training program. Following a review of some specific measures of safety, selected statistical techniques that have found application in the analysis of safety programs are discussed. Many examples in this important area of safety training are offered.

### MEASURES OF SAFETY

#### Accident Frequency Rate

The accident frequency rate (AFR), also known as the accident rate, is the number of disabling (lost-time) injuries per a specific number of annual worker-hours. For many years the number of worker-hours used was one million and was based on guidance provided by the American National Standard Institute (ANSI) Standard Z16.1. The one million worker-hours standard is based on the hours of work exposure associated with 500 employees working 40 hours each week for 50 weeks each year. Early in its history OSHA, in recognition that most firms employ far fewer than 500 employees, established a standard based on 100 employees. Today accident frequency rates are calculated using $S$, the 100 employee standard of 200,000 worker-hours.

$$S = 100 \text{ employees} \times 40 \frac{\text{hours}}{\text{weeks}} \times 50 \frac{\text{weeks}}{\text{year}}$$

$$= 200{,}000 \text{ worker-hours}$$

For a firm to determine its own AFR, it need only divide the number of lost-time injuries recorded for a given 1-year period (not necessarily a calendar year) by the actual number of worker-hours for that period, and multiply the result by 200,000.

The mathematical model recommended for use by data analysts is

$$AFR = \frac{DI}{(EMP)(AH)} \times S \tag{13.1}$$

where $DI$ = number of disabling injuries that occurred during any 12-month period

$EMP$ = number of employees at a given operational site(s), or entire organization

$AH$ = annual hours worked by one employee

= (40)(50) = 2,000 worker-hours

$S$ = 200,000 worker-hours.

The accident frequency rate is an important value primarily because of its widespread use. Many firms use the AFR to compare the results of their accident history to national averages. These figures are available for all businesses, for all industrial firms, for specific standard industrial codes (SICs), and for various levels of organizational size.

***Example 13.1***

The Uncommon Medals Company (UMC) has 30 employees at its Oakland facility and 500 employees throughout its organization distributed over 15 locations. In 1993 the Oakland Division recorded 5 disabling injuries, and the company overall experienced 60 lost-time injuries. The 1993 national figures for all manufacturing firms in the medals business was 10.0 and varied from a low of 8.0 to a high of 14.0, depending on the size of the firm. Compare UMC to the national results as one way of checking the company's safety training effectiveness.

*Solution*

$$AFR \text{ (for Oakland)} = \frac{5}{(30)(40)(50)} \times 200{,}000$$

$$= 16.67 \quad \text{lost-time injuries per OSHA standard for 100 employees}$$

$$AFR \text{ (for UMC)} = \frac{60}{(500)(40)(50)} \times 200{,}000$$

$$= 12.00 \quad \text{lost-time injuries per OSHA standard for 100 employees}$$

Comparing Oakland's AFR to UMC's overall AFR, it is readily apparent that Oakland has experienced 38.9% more disabling injuries than the company as a whole.

$$\frac{16.67 - 12.00}{12.00} \times 100 = 38.9\%$$

Furthermore, UMC's AFR is 20.0% higher than the national SIC average for UMC's classification.

$$\frac{12.00 - 10.00}{10.00} \times 100 = 20.0\%$$

Since the statement of the problem does not provide employment levels for the given AFRs, further quantitative analysis cannot be performed.

Review of the given and calculated data indicates that UMC has a potentially undesirable situation if its accident frequency rate statistics are a reliable indicator of the company's safety status. Additionally, there is no question that the Oakland division has experienced a bad year and should receive further examination to ascertain the likelihood of a repeat performance.

More precise judgments that evaluate the company's overall and specific divisional performances can be made using *Accident Facts*, published annually by the National Safety Council.

### Accident Severity Rate

The accident severity rate (ASR) is the number of days lost as a result of accidents per a specific number of annual worker-hours. As with the accident frequency rate, the standard for the annual number of worker-hours has dropped from one million to 200,000 since the advent of OSHA. Like the accident frequency rate, the accident severity rate is also important because of its widespread use and it, too, is included in the annual *Accident Facts* booklet.

#### *Example 13.2*

The Uncommon Medals Company of Example 7.1 has recorded a total of 125 lost work days for its Oakland Division in 1993 and 1200 for the firm overall. The national average for firms with the applicable SIC was 200. Compare the UMC experience both internally and relative to national statistics.

*Solution*

$$AFR \text{ (for Oakland)} = \frac{125}{(30)(40)(50)} \times 200{,}000$$

$$= 416.67 \quad \text{lost-time injuries per OSHA standard for 100 employees}$$

$$AFR \text{ (for UMC)} = \frac{1200}{(500)(40)(50)} \times 200{,}000$$

$$= 240.00 \quad \text{lost-time injuries per OSHA standard for 100 employees}$$

Comparing Oakland's ASR to UMC's overall ASR, it is clear that Oakland has experienced 73.6% more lost days than the company as a whole.

$$\frac{416.67 - 240.0}{240.0} \times 100 = 73.6\%$$

Additionally, UMC's ASR is 20.0% higher than the national SIC average for UMC business classification.

$$\frac{240.0 - 200.00}{200.0} \times 100 = 20.0\%$$

The foregoing quantitative analysis suggests that Oakland has a serious problem. Its accident severity rate is nearly 75% higher than the Uncommon Medals Company overall. This, in combination with its 38.9% higher accident frequency rate, should give UMC management cause for concern. When the company's ASR is coupled with its AFR and compared to national statistics for firms in a similar business, it would be reasonable to conclude that UMC overall is not doing well relative to its SIC and that Oakland, specifically, is a leading indicator of what could happen to the entire company if necessary changes are not made soon.

### Frequency–Severity Indicator

Alone, neither the AFR nor the ASR is sufficient to provide an overall indicator of a company's safety performance. There is, however, a way to bring the two together to establish a more valid basis for evaluation. The frequency-severity indicator (FSI) is a single value resulting from appropriately combining the AFR and the ASR. The FSI is calculated using the following equation:

$$FSI = \sqrt{\frac{AFR \times ASR}{1000}} \tag{13.2}$$

***Example 13.3***
Continuing the Uncommon Medals Company Example, calculate the FSI for both the Oakland Division and for the whole company. Compare the results of these calculations to the FSI values for UMC and Oakland in 1991–1993.

| | FSI | |
|---|---|---|
| | Uncommon Medals | Oakland Division |
| AFR | | |
| 1991 | 11.50 | 12.50 |
| 1992 | 13.07 | 17.02 |
| 1993 | 12.00 | 16.67 |
| ASR | | |
| 1991 | 200.0 | 350.0 |
| 1992 | 245.0 | 405.0 |
| 1993 | 240.0 | 416.67 |

*Solution*

$$FSI\ (1991,\ \text{UMC}) = \sqrt{\frac{(11.50)(200.0)}{1000}} = 1.52$$

$$FSI\ (1992,\ \text{UMC}) = \sqrt{\frac{(13.07)(245.0)}{1000}} = 1.79$$

$$FSI\ (1993,\ \text{UMC}) = \sqrt{\frac{(12.00)(240.0)}{1000}} = 1.70$$

$$FSI\ (1991,\ \text{OD}) = \sqrt{\frac{(12.50)(350.0)}{1000}} = 2.09$$

$$FSI\ (1992,\ \text{OD}) = \sqrt{\frac{(17.02)(405.0)}{1000}} = 2.63$$

$$FSI\ (1993,\ \text{OD}) = \sqrt{\frac{(16.67)(416.67)}{1000}} = 2.64$$

$$FSI\ (1993,\ \text{Nat'l}) = \sqrt{\frac{(10.00)(200.0)}{1000}} = 1.41$$

Now management has a firmer basis for comparing its safety performance both internally and externally. It is clear that the company overall FSI has been consistently higher than that of comparable firms for the last 3 years. Furthermore, the Oakland Division's FSI has been regularly much higher than the company overall. The average FSI for UMC for the 3-year period 1991–1993 is 1.67

$$[(1.52 + 1.79 = 1.70) + 3 = 1.67]$$

whereas that for the Oakland Division for the same period is 2.45

$$[(2.09 + 2.63 = 2.64) + 3 = 2.45]$$

It can be safely assumed that the 1993 national FSI for the UMC standard industrial code is approximately what a 1991–1993 average would be (i.e., 1.41). Therefore it can be noted that UMC is 18.4% higher than the national average and that the Oakland Division is 46.7% higher than UMC.

$$\frac{1.67 - 1.41}{1.41} \times 100 = 18.4\%$$

$$\frac{2.45 - 1.67}{1.67} \times 100 = 46.7\%$$

It can be reasonably concluded that some remedial action is required by UMC management to reverse this undesirable situation.

### Average Cost per Injury

The average cost per injury (ACI) combines a number of related factors into a single value, which can then be compared to other ACIs. Costs that are usually considered in developing an ACI value for a business include actual compensation paid to an employee during the calendar period in which he or she is incapacitated and cannot work as a result of a job-related injury, medical costs (hospital, doctor, nurse, first aid, ambulance, etc.) paid for by the business in support of job-related injury, workmen's compensation premiums to the firm's insurance carrier, and others. Comparison of ACIs between departments, from year to year, and so on, provides still another measure of safety.

#### *Example 13.4*

| 1993 Injury Expenses | Dept. A | Dept. B | Dept. C | SDD Overall |
|---|---|---|---|---|
| Compensation | $10,000 | $12,000 | $ 6,000 | $28,000 |
| Medical costs | 7,000 | 14,000 | 10,000 | 31,000 |
| Insurance premiums (prorated by number of employees) | 5,000 | 6,000 | 5,000 | 16,000 |
| Miscellaneous costs | 2,000 | 3,000 | 3,000 | 8,000 |
| Total | $24,000 | $35,000 | $24,000 | $83,000 |
| 1993 injuries | 50 | 75 | 35 | 160 |
| 1993 ACI | $480 | $467 | $686 | $519 |

Analysis of the calculations indicates that there is little difference between Departments A and B; however, Department C has an ACI 32.2% higher than the overall ACI and 44.9% higher than the average of Departments A and B.

$$\frac{686 - 519}{519.00} \times 100 = 32.2\%$$

$$\frac{686 - (480 + 467)/2}{(480 + 467)/2} \times 100 = 44.9\%$$

Extending this analysis, management has requested the 1993 ACI be compared to those of previous years. Caution must be exercised when performing year-to-year comparisons that involve dollars. It is necessary to account for inflation to avoid comparison of dollars that are dissimilar in value. Continuing this example, suppose that the following company data are located in the UMC files, as well as associated historical inflation figures gleaned from government sources:

| Year | Total Injury-Related Costs | Number of Injuries | Constant $ Value | ACI |
|---|---|---|---|---|
| 1991 | $75,000 | 150 | $1.15 | $575 |
| 1992 | $74,000 | 152 | $1.08 | $526 |
| 1993 | $83,000 | 160 | $1.00 | $519 |

where

$$ACI \text{ (for 1991)} = \frac{(75{,}000)(1.15)}{150} = \$575$$

$$ACI \text{ (for 1992)} = \frac{(74{,}000)(1.08)}{152} = \$526$$

$$ACI \text{ (for 1991)} = \frac{(83{,}000)(1.00)}{160} = \$519$$

This analysis indicates that the average cost per injury has decreased 9.74% from 1991 to 1993.

$$\frac{575 - 519}{575} \times 100 = 9.74\%$$

Had management ignored the impact of inflation in its analysis, the ACI for 1991, calculated with the 1991 value dollar, would have been $75,000 + 150 = $500. This inaccurate analysis would have indicated that the ACI had increased by 3.80% from 1991 to 1993.

$$\frac{519 - 500}{500} = 3.80\%$$

Only when the 1993 value of the 1991 dollar is used, as in the preceding tabulation, is an inflation-adjusted figure obtained.

### Safe-T-Score

Up to this point, the measures of safety presented have made use of simple, basic math. The Safe-T-Score technique, which employs the results of accident frequency rate calculations, goes beyond basic math. It makes use of a statistical analysis technique referred to as the Student's *t*.

$$Safe\text{-}T\text{-}Score = \frac{AFR\ (\text{now}) - AFR\ (\text{past})}{\sqrt{\dfrac{AFR\ (\text{past})}{\text{worker-hours (now)}/200{,}000}}} \tag{13.3}$$

Before demonstrating its use through an example, certain characteristics of Safe-T-Score are worth considering. A Safe-T-Score value does not have any dimensions; that is, it is an artificially created number that is not descriptive of a physical characteristic such as cost (dollars), weight (pounds), or area (square feet). Only the algebraic sign and the value of the Safe-T-Score are of interest to the analyst; a positive sign indicates a worsening situation, and a negative value reflects an improvement of the present situation over the past.

With reference to Figure 13.1, the value of the Safe-T-Score is descriptive of certain conditions as presented in Table 13.1.

#### *Example 13.4*

Once again the Uncommon Medals Company is desirous of reviewing its safety performance data to ascertain the status of its safety program. This time UMC has decided to use the Safe-T-Score approach to analyze its interpersonal data. The following 1991–1993 information has been collected for use by an analyst (from Example 13.3).

| Data | Oakland Division | UMC (Overall) |
|---|---|---|
| 1992 | | |
| Accidents | 4 | 49 |
| Worker-hours | 47,000 | 750,000 |
| AFR | 17.02 | 13.07 |
| 1993 | | |
| Accidents | 5 | 60 |
| Worker-hours | 60,000 | 1,000,000 |
| AFR | 16.67 | 12.00 |

The AFR for the Oakland division has decreased by 2.06% from 17.02 to 16.67, whereas the AFR for the entire company has decreased by 8.19% from

13.07 to 12.00. Without the use of Safe-T-Score it cannot be said with certainty whether an event of any statistical significance has occurred. Let us examine the solution.

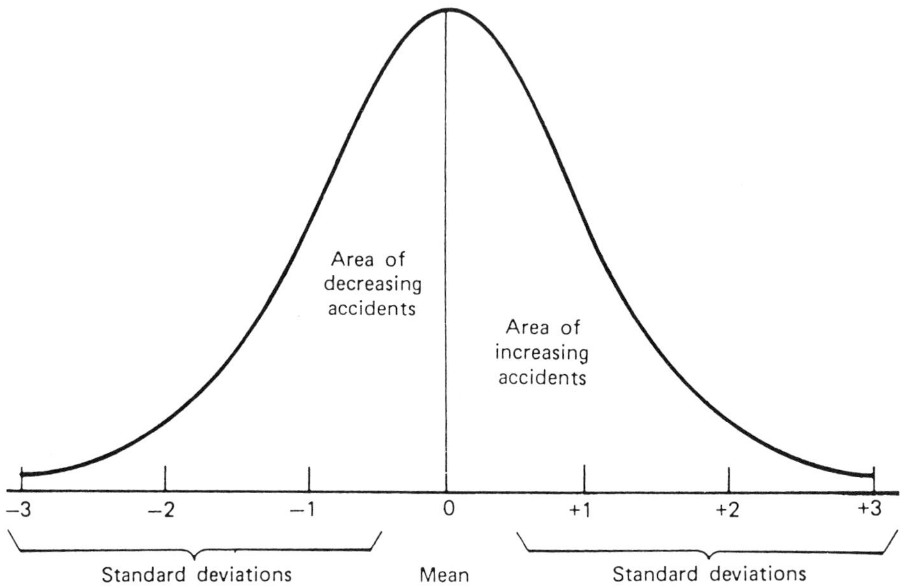


**FIGURE 13.1** Safe-T-Score profile.

**TABLE 13.1 Safe-T-Score Interpretations**

| Safe-T-Score | Interpretation |
|---|---|
| Between +3.00 and −3.00 standard deviations | AFR changes are not significantly different; variation is due to chance |
| Over +3.00 standard deviations | AFR is now significantly worse than it was in the past; investigation of changes are necessary |
| Under −3.00 standard deviations | AFR is now significantly better than it was in the past; considerably fewer accidents are being reported |

*Solution.* Using Eq. (13.3), the Safe-T-Score for the Oakland division is calculated as follows:

$$\textit{Safe-T-Score}\ (\text{for Oakland}) = \frac{16.67 - 17.02}{\sqrt{\dfrac{17.02}{60{,}000/200{,}000}}}$$

$$= \frac{-0.35}{\sqrt{17.02/0.30}} = \frac{-0.35}{\sqrt{56.73}}$$

$$= \frac{-0.35}{7.53} = -0.046$$

Since this Safe-T-Score is very nearly zero, it is determined that virtually no change has taken place.

Again using Eq. (13.3), the Safe-T-Score for the entire Uncommon Medals Company is determined:

$$\textit{Safe-T-Score}\ (\text{for UMC}) = \frac{12.00 - 13.07}{\sqrt{\dfrac{13.07}{1{,}000{,}000/200{,}000}}}$$

$$= \frac{-1.07}{\sqrt{13.07/5.00}} = \frac{-1.07}{\sqrt{2.61}}$$

$$= \frac{-1.07}{1.62} = -0.66$$

Since this Save-T-Score is between zero and −3.00, it is reasonable to believe that nothing of any significance has occurred; if something *has* happened, however, we can assume that it is for the better.

## Attitude Survey

Elsewhere is this book it is recommended that an annual employee attitude survey be distributed, collected, and analyzed to assess the effectiveness of a company safety program. Regardless of the questions that are included in the instrument, two characteristics of the survey are valuable tools in the evaluation of a safety program.

#### *Attitude Survey Response Rate*

The level of response to an annual survey will vary considerably between companies as well as within a company from year to year. If all other variables are held constant, the survey response rate becomes a thermometer directly measuring the employees' interest and belief in and support of the company safety program. Some of the other variables that must be held as nearly constant as possible include number of questions in the survey, question content, methods of distribution and collection, and time of the year (relative to contract negotiations, payday, holiday, etc.).

### Example 13.6

The Uncommon Medals Company annually asks both its hourly and salaried employees to complete an attitude survey instrument that is oriented primarily toward the company safety program. Each year the survey is distributed and collected in a consistent manner, with only minimal changes in question wording to reflect recent changes in the safety program.

The UMC director of safety has been keeping track of the survey results for the five years since it began, as follows:

| Year | Number Distributed | Number Returned | Number Usable | Response Rate (%) | Usable Rate (%) |
|---|---|---|---|---|---|
| 1989 | 144 | 72 | 65 | 50 | 90 |
| 1990 | 150 | 84 | 78 | 56 | 93 |
| 1991 | 158 | 104 | 98 | 66 | 94 |
| 1992 | 180 | 108 | 92 | 60 | 85 |
| 1993 | 200 | 146 | 139 | 73 | 95 |

Attempt to relate these results to the company's employee history during this period.

*Solution.* In 1989 only 50% of the employees returned the survey. In retrospect it appears that this low point represents an attempt by the workforce to understand what management was up to with this new feature of the company safety program.

In 1990 and 1991 both the response rate and the usable rate grew dramatically as the employees became more comfortable and more familiar with the instrument.

In 1992, however, these two indicators dropped significantly; a portent of trouble about to surface. Just 60 days after the survey was conducted, union problems nearly precipitated a strike that was narrowly averted. It is possible that UMC management first became aware of the potential problem through a knowledge of the survey results.

In 1993 things appear to have returned to normal with both rates back on their original tracks. Any negative deviation of these trend lines in the future can be used to predict present or developing problems.

### *Attitude Survey Content Variance*

The extent of dispersion surrounding the mean response to a specific question can be a "red flag" indicating considerable variation of opinion regarding the question's content. Minimal variation indicates that there is little difference of opinion about the topic addressed. On the other hand, a high level of dispersion about the mean should be interpreted as an indicator that a consensus does not exist. Whether dispersion is high or low may be difficult to ascertain the first year a survey is conducted. In succeeding years, however, the expanded data base should provide the survey analyst with sufficient background information to make these judgments.

### *Example 13.7*

Continuing the analysis of the Uncommon Medals Company annual employee attitude survey, the following partial results have been documented.

| | Year | | | | |
|---|---|---|---|---|---|
| Question No.[a] | 1989 | 1990 | 1991 | 1992 | 1993 |
| 1. Mean[b] | 3.5 | 3.6 | 3.7 | 3.6 | 3.7 |
| SD[c] | 0.4 | 0.4 | 0.3 | 0.3 | 0.3 |
| 2. Mean | 2.4 | 2.2 | 2.4 | 2.5 | 2.3 |
| SD | 0.5 | 0.7 | 0.7 | 0.8 | 1.0 |
| 3. Mean | 2.1 | 2.0 | 2.0 | 2.7 | 2.8 |
| SD | 0.8 | 0.6 | 0.6 | 0.5 | 0.4 |
| 4. Mean | 4.0 | 3.7 | 3.2 | 3.0 | 3.0 |
| SD | 0.3 | 0.4 | 0.6 | 0.8 | 0.9 |
| 5. Mean | 3.0 | 3.2 | 3.5 | 3.8 | 4.0 |
| SD | 0.5 | 0.5 | 0.4 | 0.3 | 0.4 |

[a] The responses to each question are based on a scale of 1 (low) to 5 (high).
[b] The mean (also known as the arithmetic mean, or average), a measure of central tendency.
[c] The standard deviation (SD), a statistical measure of dispersion of data around the arithmetic mean.

Review these results in an attempt to identify potential areas of concern.

*Solution.* Reviewing the data available on a year-by-year basis reveals some important highs and lows in the mean responses to the various questions, but it does not offer much assistance in identifying safety-related personnel problems. A longitudinal analysis, however, offers considerably more insight on several counts.

Question 1 shows high year-to-year consistency in both mean and standard deviation. Question 2, on the other hand, shows a similar level of consistency in its recorded means, but over the 5-year period the standard deviation has gradually doubled. This indication of considerable change in the opinions of the respondents serves to identify a potential area for investigation.

Question 3 shows a 33% increase in the mean and a 50% decrease in the standard deviation between 1989 and 1993. Clearly the employees' attitudes on this topic are following an upward trend and, simultaneously, becoming more consistent (less content variance).

Question 4 reflects a situation quite the opposite of Question 3. The mean has dropped by 25% and the variance has increased by a factor of 3. This dramatic change should be cause for concern, and it justifies a thorough investigation of the cause(s).

Finally, the mean of Question 5 reflects a 33% increase (equivalent to Question 3), but the standard deviation is relatively unchanged. This result could be considered to represent an improving trend with little content variance.

The foregoing analysis is a simple example demonstrating the type of information and indicators that are available in reviewing periodic employee attitude surveys beyond the original information expected from the questions.

### Hazard Reporting Rate

The hazard reporting rate (HRR) is a function of at lease five contributing factors: number of physical hazards that require reporting, number of personnel hazards that require reporting, motivation of personnel to report hazards that have been detected, capacity of personnel to discern the presence of hazards, and the organizational climate (which may or may not be supportive of a high HRR).

The number of physical hazards that require reporting depends primarily on what fraction of a firm's physical hazards has already been reported and the extent of change that occurs within the organization's facilities and equipment. A discussion of physical hazards is found in Chapter 8.

The number of personnel hazards that require reporting is directly related to the extend to which a firm's employees perform their jobs responsibly. When safe job performance is not considered to be paramount, the number of personnel hazards will be excessive.

The level of employees' motivation to report hazards that have been detected is frequently associated with the desirability of the reward system provided by employers and as perceived by the employees. A firm may believe that its reward system is ideally suited to motivate the employees to report hazards, but unless the system is viewed in a similar fashion by the employees, it is unlikely that the employees will be moved to search out and report hazards.

The capacity of personnel to discern the presence of hazards is closely linked to their ability to recognize hazards. This requires considerable attention to training, since most employees are not tuned into identifying hazards that may be in daily proximity to their work areas. A training program especially tailored to upgrading employee hazard recognition skills must be a regular part of a company safety training program.

The organizational climate may or may not be supportive of a high HRR. This factor reflects the extent to which employees in general are cooperating with management to make a firm a safer place to work. When an employee identifies a hazard but neglects to report it despite a desire to receive a reward provided by the employer, it must be assumed that the employee fears condemnation or some other form of mental or even physical punishment by fellow employees. When this is the case, management should be concerned about more than just the reporting hazards.

Hazard reporting rates can be recorded for analysis in a variety of different formats. Example 7.8 offers some ideas on this subject.

***Example 13.8***

The director of safety of the Uncommon Medals Company has been keeping records on hazard reporting for 3 years. Now that a sufficient data base exists, the director has begun to consider some ideas on presentation, to facilitate whatever analysis may eventually be required. Accept the responsibility for this project and offer some tabular methods for data presentation.

*Solution.* One way to present hazard reporting data is on a gross overall basis as in Table 13.2. This method is easily subdivided into weekly, monthly, or quarterly results, as in Table 13.3.

Another way to look at these figures is by type of hazard, that is, physical or personnel, as in Table 13.4.

**TABLE 13.2 Overall Approach to Hazard Reporting**

| Year | Number of Hazards Reported |
|---|---|
| 1991 | 84 |
| 1992 | 96 |
| 1993 | 120 |

**TABLE 13.3 Periodic Approach to Hazard Reporting**

| | Number of Hazards Report | | | | |
|---|---|---|---|---|---|
| Year | 1st Quarter | 2nd Quarter | 3rd Quarter | 4th Quarter | Total |
| 1991 | 20 | 20 | 24 | 20 | 84 |
| 1992 | 25 | 30 | 35 | 6 | 96 |
| 1993 | 30 | 40 | 35 | 15 | 120 |

**TABLE 13.4 Categorical Approach to Hazard Reporting**

| | Type of Hazard | | |
|---|---|---|---|
| Year | Physical | Personnel | Total |
| 1991 | 48 | 35 | 48 |
| 1992 | 58 | 38 | 96 |
| 1993 | 84 | 36 | 120 |

If the director of safety has reason to be concerned about a specific department's lack of involvement in the hazard reporting system, an arrangement such as that shown in Table 13.5 could be employed.

Naturally, any of these approaches can be combined to offer greater insight on the hazard reporting system. For example, combining the information contained in Tables 13.4 and 13.5 could result in either Table 13.6A or 13.6B, depending on the director's need and/or analysis orientation.

If the departmental sizes are approximately equal, presentations such as the foregoing may suffice to provide usable data for analysis. On the other hand, if the departments are of significantly different sizes, or perhaps the sizes are changing from one reporting period to another, it is sometimes useful to put the data on a per capita basis. If the departmental sizes were as described in Table 13.7, the information contained in Table 13.5 would appear as in Table 13.8.

The foregoing should not be received as a complete discussion, but rather a beginning approach to hazard reporting analysis.

## Facilities Appearance; Housekeeping

Still another measure of effectiveness of a company safety training program is the appearance of the company facilities and equipment; this is frequently referred to as housekeeping.

To be able to evaluate the status of housekeeping within a specific area, a baseline of acceptable attributes must be established (Table 13.9). Once this checklist has been carefully developed, it can serve as a basis for all future evaluations.

**TABLE 13.5 Departmental Approach to Hazard Reporting**

| | Department of Reporter | | | |
|---|---|---|---|---|
| Year | Dept. X | Dept. Y | Dept. Z | Total |
| 1991 | 15 | 25 | 44 | 84 |
| 1992 | 30 | 20 | 46 | 96 |
| 1993 | 45 | 30 | 45 | 120 |

**TABLE 13.6A Combined Approach to Hazard Reporting: By Department**

| | Department of Reporter | | | | | | | | |
|---|---|---|---|---|---|---|---|---|---|
| | Dept. X | | Dept. Y | | Dept. Z | | Total | | |
| Year | Physical Hazard | Personnel Hazard | Physical Hazard | Personnel Hazard | Physical Hazard | Personnel Hazard | Physical Hazard | Personnel Hazard | Total |
| 1991 | 10 | 5 | 15 | 10 | 23 | 21 | 48 | 36 | 84 |
| 1992 | 20 | 10 | 10 | 10 | 28 | 18 | 58 | 38 | 96 |
| 1993 | 25 | 20 | 10 | 20 | 25 | 20 | 84 | 36 | 120 |

**TABLE 13.6B Combined Approach to Hazard Reporting: By Type**

| | Type of Hazard | | | | | | | | |
|---|---|---|---|---|---|---|---|---|---|
| | Physical Hazard | | | Personnel Hazard | | | Total | | |
| Year | Dept. X | Dept. Y | Dept. Z | Dept. X | Dept. Y | Dept. Z | Dept. X | Dept. Y | Dept. Z |
| Total | | | | | | | | | |
| 1991 | 10 | 15 | 23 | 5 | 10 | 21 | 15 | 25 | 44 | 84 |
| 1992 | 20 | 10 | 28 | 10 | 10 | 18 | 30 | 20 | 46 | 96 |
| 1993 | 25 | 10 | 25 | 20 | 20 | 20 | 45 | 30 | 45 | 120 |

**TABLE 13.7 Department Sizes**

| | Number of Employees | | | |
|---|---|---|---|---|
| Year | Dept. X | Dept. Y | Dept. Z | Total |
| 1991 | 50 | 75 | 125 | 250 |
| 1992 | 75 | 100 | 125 | 300 |
| 1993 | 100 | 100 | 100 | 300 |

**TABLE 13.8 Per Capita/Departmental Approach to Hazard Reporting[a]**

| | Department of Reporter | | | |
|---|---|---|---|---|
| Year | Dept. X | Dept. Y | Dept. Z | Total |
| 1991 | 0.30 | 0.33 | 0.35 | 0.34 |
| 1992 | 0.40 | 0.20 | 0.37 | 0.32 |
| 1993 | 0.45 | 0.30 | 0.45 | 0.40 |
| Average | 0.40 | 0.27 | 0.39 | 0.35 |

[a]Annual reports of hazards per employee

**TABLE 13.9 Customized Housekeeping Checklist**

Department X
Month October, 1993

| | Date | | | | | | | | | |
|---|---|---|---|---|---|---|---|---|---|---|
| Item | 1 | 2 | 3 | 4 | 5 | ........ | 28 | 29 | 30 | 31 |
| 1. Portable electrical equipment is grounded | | | | | | | | | | |
| 2. "No Smoking" signs are posted and observed | | | | | | | | | | |
| 3. Fire aisles are kept clear | | | | | | | | | | |
| . | | | | | | | | | | |
| . | | | | | | | | | | |
| . | | | | | | | | | | |
| 9. First aid kit contents are fresh and complete | | | | | | | | | | |
| 10. Hard hats are in use by all personnel in area | | | | | | | | | | |
| Overall | | | | | | | | | | |

***Example 13.9***

The Uncommon Medals Company has always emphasized the importance of good housekeeping to the company safety program. After many hours of research the director of safety has produced a new, more comprehensive series of housekeeping checklists for use by persons conducting inspections of specific departments. Each department and specialized work area has its own customized checklist, which is designed to ensure the fairest and most thorough appraisal of housekeeping possible. Produce a customized housekeeping checklist for Department X.

*Solution.* In Table 13.9, the customized housekeeping checklist for use in Department X, each item should be evaluated and a value assigned for that item for that day. Values can be assigned in any number of ways depending on local preferences.

Table 13.10 offers some ideas for quantitative evaluatory systems for use in conjunction with a housekeeping checklist such as Table 13.9.

### Other Measures

Additional measures of safety training effectiveness that may be considered include statistical control charts, workmen's compensation premium ratings, personnel productivity indicators, personnel evaluation ratings, and safety performance indicators. Numerous authoritative references are to be found in the Appendix. These references are logical points to being a continuing search for other measures that may have application in your organization.

## STATISTICAL TECHNIQUES

### Correlation Analysis

Numerous opportunities to use correlation analysis exist in the safety training area. A discussion of this widely used statistical analysis technique can be located in nearly any introductory statistics text.

**TABLE 13.10 Quantitative Grading Systems for Use with Checklists**

| | Quantitative System | | | | | |
|---|---|---|---|---|---|---|
| Evaluation | I | II | III | IV | V | VI |
| Excellent | +2 | 5 | 4 | +3 | 8 | 10 |
| Good | +1 | 4 | 3 | +1 | 6 | 8 |
| Average | 0 | 3 | 2 | 0 | 4 | 6 |
| Fair | −1 | 2 | 1 | −1 | 2 | 4 |
| Poor | −2 | 1 | 0 | −3 | 0 | 2 |

Some of the areas in which correlation analysis has been successfully used include

1. Training program examination scores versus personnel evaluation ratings
2. Training Program examination score (average) for a department or shift versus the departmental or shift accident rate
3. Training program delta scores (the difference between a training program pretest and the final exam) versus personnel evaluation ratings
4. Training program delta score (average) for a department or shift versus the departmental or shift accident rate

Example 13.10 demonstrates the use of correlation analysis in the comparison of training program examination scores and personnel evaluation ratings.

***Example 13.10***

The director of safety of the Uncommon Medals Company is interested in the possible linear relationship between an employee's score on an awareness safety knowledge (ASK) examination and that employee's safety rating by his or her supervisor at the completion of the first year of employment. A random sample of seven employees provided the results presented in Table 13.11. Let $X$ be the ASK examination score, and let $Y$ be the supervisor's rating. In each case the maximum possible score ($X$) or rating ($Y$) is 15. The director of safety wishes to know whether there is any apparent relationship between $X$ and $Y$, and if such exists, whether it is linear or nonlinear.

*Solution.* First, as is appropriate in most statistically oriented problems, we construct a diagram. In this case we use a scatter diagram of the ASK scores on the horizontal ($X$) axis and of the supervisor's ratings on the vertical ($Y$) axis (Figure 13.2).

A visual review indicates that a relationship exists; that is, in general as $X$ values increase, $Y$ values also increase. It is not clear from a visual inspection whether a linear relationship exists between the two variables. To determine whether such a relationship actually exists, it is necessary to compute the correlation coefficient $r$. The mathematical model for $r$ is as follows:

$$r = \frac{n\Sigma XY - \Sigma X \Sigma Y}{\sqrt{n\Sigma X^2 - (\Sigma X)^2} \cdot \sqrt{n\Sigma X^2 - (\Sigma X)^2}} \tag{13.4}$$

where

$n$ = the number of data pairs

$\Sigma$ = sigma (uppercase Greek), which indicates a requirement to sum all values

$\Sigma XY$ = the sum of all the products resulting from multiplying each $XY$

$\Sigma X$ = the sum of all the $X$ values

$\Sigma Y$ = the sum of all the $Y$ values

$\Sigma X^2$ = the sum of the squares of all the $X$ values

$\Sigma Y^2$ = the sum of the squares of all the $Y$ values

$(\Sigma X)^2$ = the squared value of $\Sigma X$

$(\Sigma Y)^2$ = the squared value of $\Sigma Y$

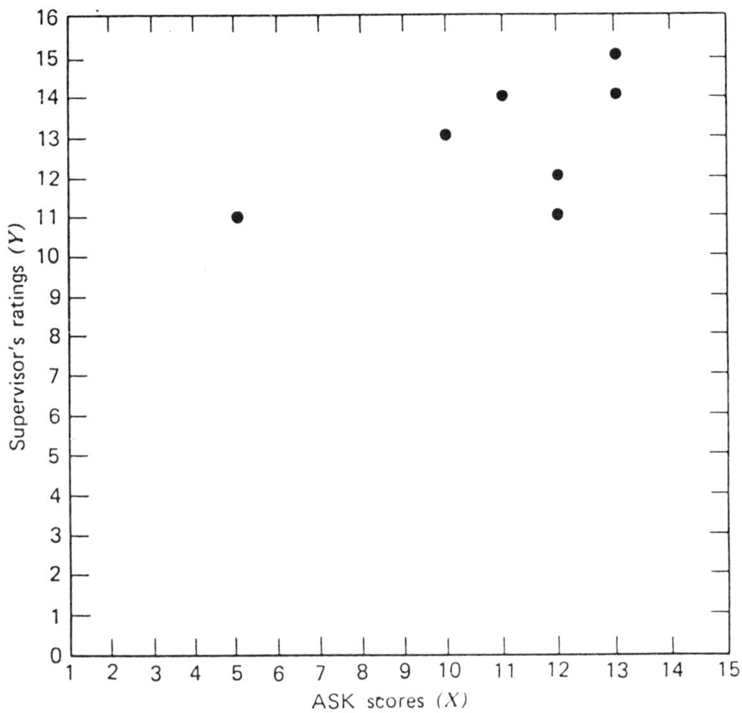


**FIGURE 13.2** Scatter diagram for Example 13.10.

To obtain these values for use in computing $r$, it is necessary to expand Table 13.11 to show the required calculations. This expansion is presented in Table 13.12.

For Employee No. 1, whose $X$ value was 12 whose $Y$ value was 11, the product of multiplying 12 times 11 is 123, the $XY$ value. The square of $X$ (i.e., 12) is 122 and of $Y$ is 121. These data are carried out for each employee.

When the expansion is complete, each column is summed, to produce the following results:

$$
\begin{aligned}
n &= 7 \text{ (the number of data pairs)} \\
\Sigma XY &= 992 \\
\Sigma X &= 76 \\
\Sigma Y &= 90 \\
\Sigma X^2 &= 872 \\
\Sigma Y^2 &= 1172 \\
(\Sigma X)^2 &= (76)^2 = 5776 \\
(\Sigma Y)^2 &= (90)^2 = 8100
\end{aligned}
$$

When applied to the mathematical model for $r$, these values generate the value of $r$.

$$
\begin{aligned}
r &= \frac{7(992) - (76)(90)}{\sqrt{7(872) - 5776} \cdot \sqrt{7(1172) - 8100}} \\
&= \frac{6944 - 6840}{\sqrt{6104 - 5776} \cdot \sqrt{8204 - 8100}} \\
&= \frac{104}{\sqrt{328} \cdot \sqrt{104}} \\
&= \frac{104}{(18.111)(10.198)} \\
&= \frac{104}{184.694} \\
&= +0.563
\end{aligned}
$$

This value of $r$, the correlation coefficient, measures the strength of the linear relationship between $X$ and $Y$.

Since values of $r$ can vary from zero to either plus or minus ($\pm 1$), this analysis indicates nearly 60% positive correlation in favor of a linear relationship between the ASK test scores and supervisory ratings of employees' safety-related performance. Based on this information, the UMC safety director will very likely decide to continue to use the ASK exam, but will certainly attempt to improve it.

**TABLE 13.11 ASK and Rating Results**

| Employee | $X$ | $Y$ |
|---|---|---|
| 1 | 12 | 11 |
| 2 | 11 | 14 |
| 3 | 5 | 11 |
| 4 | 10 | 13 |
| 5 | 13 | 15 |
| 6 | 13 | 14 |
| 7 | 12 | 12 |

**TABLE 13.12 Example 13.10 Data Expansion**

| Employee | $X$ | $Y$ | $XY$ | $X^2$ | $Y^2$ |
|---|---|---|---|---|---|
| 1 | 12 | 11 | 132 | 144 | 121 |
| 2 | 11 | 14 | 153 | 121 | 196 |
| 3 | 5 | 11 | 55 | 25 | 121 |
| 4 | 10 | 13 | 130 | 100 | 169 |
| 5 | 13 | 15 | 195 | 169 | 225 |
| 6 | 13 | 14 | 182 | 169 | 196 |
| 7 | 12 | 12 | 144 | 144 | 144 |
| | 73 | 90 | 992 | 872 | 1172 |

## Regression Analysis

Frequent use of regression analysis in the assessment of safety training methods indicates the high regard among safety professionals for the versatility and utility of this approach.

Regression analysis has been successfully applied in numerous safety training-related situations such as the following:

1. Forecasting accident frequency and severity rates for future time periods
2. Development of mathematical instruments to predict the numerical values of future occurrences

**Scrape.** A procedure designed to provide management with periodic informational up-dates on a company's accident prevention performance.

**Stratification.** A method of categorizing data into two or more strata to determine what effect each category of data has on an independent–dependent variable relationship, e.g., in a **Scatter Analysis** of height and weight, men generally weigh more than women for a given height.

**Tally Sheets.** A graphic approach to data collection where specific categories (using either attributes such as A, B, C or variables such as inclusive ranges of dimensions, weights, angles) are established, followed by counts of occurrences using tally marks to facilitate data recording, e.g., 𝍸 [see *Checklists/Sheets*].

**Tree Diagram.** A graphic tool used to systematically and continually develop ideas or issues, and then expand them in greater detail. Used as a planning tool, it can reveal all necessary means and activities to successfully implement a program.

## ADDITIONAL INFORMATION

In addition to the information presented so far, OSHA Pamphlet 7, "Safety and Health Quality Assurance," also provides valuable guidance on evaluating safety programs. It is reproduced as follows:

**Safety and Health Quality Assurance**
**Determining How Well Your Safety and Health Program Works**

### INTRODUCTION

You know your safety and health program is in place. You have set your safety and health goal for the year and clearly stated the objectives, procedures and activities necessary to meet that goal. Responsibilities have been defined and clearly assigned, and adequate authority and resources have been allocated to carry out the responsibilities. People have been trained in their roles in the safety and health program, and they clearly know the consequences if they do not carry out their assigned responsibilities.

But your responsibility for employee safety and health is not fully discharged. The next critical step is to evaluate how well your safety and health program is working to achieve your goal. This process is not an inspection. Inspections are necessary to look at the facility, the process, and each job to see if any hazards exist so that they may be eliminated or controlled.

This process is also more than the usual audit. For example, if you are evaluating employee participation by looking at the activities of the safety committee, you will want to know if that committee met at the intervals specified, and if most of the members attended each meeting. These are audit questions. But you want to move beyond this simple accounting to ask if and how employee participation at these meetings helped improve the safety and health program at your site. How is the work of the safety committee helping you meet your goal? These are evaluation issues. Interviewing the committee members might lead you to some answers. Asking supervisors whether activities of the committee have changed employee behavior or changed attitudes in the workplace such that safety and health are improved will also help the evaluator judge the committee's effectiveness.

A safety and health evaluation looks at the systems you have created to carry out your safety and health program to see if the systems are working efficiently to enable you to reach your goal.

Three tools are useful for this evaluation: (1) document review, (2) employee interviews, (3) review of site conditions. Although evaluations can be done by persons at the site, the best evaluations are done by people who are knowledgeable about the process(es) at the site and about managing safety and health programs, but who are not working at the site being evaluated. The fresh look an outsider brings produces a more accurate and helpful evaluation. This outsider may come from corporate headquarters or another worksite within the company, an insurance company, or a consulting firm.

All systems that contribute to your safety and health program should be reviewed. These should include management leadership and the evaluation of that leadership, the analysis of your worksite to identify hazards, hazard prevention and control, accident and near-miss investigations, employee involvement, safety and health training, use of personal protection equipment, adequacy of the health program and the emergency response program. Your site may have additional programs or systems that contribute to the safety and health program that you need to evaluate.

Evaluation often causes anxiety among the people at the site to be evaluated. When planning the evaluation, if you clarify that the evaluator's focus will be on the systems rather than on the personnel, some of that anxiety may be dissipated.

The final report should be a narrative account, not a check list. Some people include pictures in the report both of situations that are excellent and those that must be improved. Some provide a grading system so that each year's results can quickly be compared to previous years. But the body of the report should list the programs or systems reviewed and a narrative report of the examination of each system or program. The report should contain a schedule for when changes will be completed, who is responsible for seeing that the changes are made, and when the changes were actually completed. This report should be available to any employee who wants to read it.

For safety and health to remain a visible top priority at any site, good management programs must be in place, but these programs should be examined yearly to be sure that they continue to contribute to meeting the goal set for that year, and to make the changes necessary to point the programs toward the goal set for the next year.

The material in this pamphlet will explain in detail, with examples, just how this examination can be effectively done.

## WHAT SHOULD BE EVALUATED?

Ideally everything that you know to be directly contributing to your safety and health program should be evaluated. Since OSHA has determined and published the elements of a good safety and health program as the OSHA Safety and Health Program Management Guidelines (54 FR 3908, January 26, 1989), the outline of that publication should help you determine areas to be covered by your evaluation. The outline is as follows:

- The demonstration of management leadership through:
  - Setting and communicating the safety and health policy;
  - Setting and communicating a clear goal and objectives;
  - Being visibly involved in employee safety and health;
  - Assuring employee involvement in safety and health problem identification and resolution;
  - Clearly assigning responsibility for safety and health;
  - Giving adequate authority and assuring efficient use of resources;
  - Holding line supervisors accountable;
  - Assuring quality.
- The analysis of your worksite to identify hazards through:
  - Comprehensive safety and health hazard surveys;
  - Analysis of planned changes to identify hazards they might introduce;
  - Routine hazard analyses such as;
    —Job hazard analysis (also known as job safety analysis);
    —Process hazards analysis (used in industries with complex and hazardous processes); and,
    —Phase hazard analysis (used in construction, mostly);
  - Employee reports of hazards;
  - Accident/incident investigations; and,
  - Analysis of injury/illness trends.
- Hazard prevention and control techniques through:
  - Engineering controls;
  - Work practice controls;
  - Personal protective equipment;
  - Administrative controls (when necessary);
  - Disciplinary systems to enforce controls;
  - Preventive maintenance;
  - Emergency preparedness; and,
  - Medical program.
- Providing safety and health training to ensure that your employees at all levels know how to protect themselves and others from the hazards and potential hazards of your worksite.

### Summary

Every aspect of your safety and health program should be evaluated against your overall goal, your objectives and the procedures you set to achieve them.

## WHO SHOULD EVALUATE?

The person or persons selected to do a safety and health program evaluation can be drawn from the workplace safety and health department or the safety committee. But the best evaluators will be people not involved with the day to day operations of the site. Sources for finding an evaluator with fresh eyes for the site are the corporate safety department, a knowledgeable person from another site in the company, insurance companies, and consultants. It is sometimes helpful to have two activity managers switch and evaluate each other's results.

The evaluator(s) should be knowledgeable in the technical aspects of occupational safety and health, managing safety and health, and evaluating programs. Of the three areas of knowledge, management of safety and health is the most important. If you want to have your evaluator(s) trained in safety and health management, you may contact the VPP Participants Association at the following address or telephone number to inquire about the course on OSHA's Safety and Health Program Management Guidelines.

VPP Participants Association
P.O. Box 991
McLean, Virginia 22101
(703)761-1146

### Summary

The evaluator(s) should be knowledgeable about safety and health in your operations and about managing safety and health programs, but should not be too close to the actual operations.

## TOOLS FOR COLLECTING INFORMATION USED IN EVALUATION

OSHA has found that there are three indispensable evaluation tools for judging the effectiveness of occupational safety and health program management. These are:

- review of written programs and records of activity;
- interviews with employees at different levels; and,
- review of site conditions.

For a detailed description of how to use these tools, see Appendix 2, "Further Description of a Safety and Health Program Assessment," on page 22 of this pamphlet.

### Documentation

Every worksite will have, at an absolute minimum, written accident reports and the OSHA log of injury and illnesses as required by law. Major companies should have written procedures and records of all their safety and health programs. The evaluator should compare the written program to the written records of what occurred.

### Interviews

In addition to the documentation or where it is not (fully) available, interviews can be very helpful as a second source to establish what has occurred. OSHA uses two kinds of interviews, called "formal" and "informal." The formal interviews are conducted with randomly selected employees who are asked preselected questions. Informal interviews occur at employee work stations and loosely follow a list of topics.

To assess how well the worksite safety and health policy is communicated and understood and how well the disciplinary system works, ask the employees to explain them.

To understand how well safety and health training has succeeded, interview hourly employees and first-line supervisors. Ask employees to describe what hazards they are exposed to, and how they are protected. Ask them to tell what they are supposed to do in several different types of emergencies. Ask supervisors how they teach, how they reinforce the teaching, how they enforce safety and health rules and safe work practices, and what their responsibilities are during emergency situations.

Interviews with management about safety and health program activities should focus on management's involvement and commitment to the safety and health program. Ask how the policy statement was created, and how that statement is communicated to all employees. Ask what information management receives about the safety and health activities and what action management takes as a result of that information. Ask how management commitment to safety and health is visible to the workforce.

### Review of Site Conditions

The conditions at the worksite reveal much about the safety and health program's effectiveness. Worksite conditions can be seen indirectly by looking at documents such as inspection reports of hazards, employee reports of hazards, and accident/incident investigations.

Site tours may also reveal hazards. Be careful, however, that the site tour does not become an inspection, with emphasis on hazard correction. When a hazard is found, in addition to assuring that it is corrected, ask what management system should have prevented or controlled the hazard. Determine why that system failed, and also correct the system. (For more information on this technique, see Pamphlet #9 in the Safety and Health Program Assistance Series [hereafter called 'the Series'], "Catching the Hazards That 'Escape' Controls," and, in particular, "Hazard Analysis Flow" which is at the back of that pamphlet.

### Summary

Three tools are especially helpful for gathering information with which to evaluate your safety and health program: review of written programs and documentation of program activity; interviews with employees at all levels; and analysis of hazards found in the workplace to see what management system failures allowed the hazard to occur.

## EVALUATE SAFETY AND HEALTH PROGRAM ACTIVITIES TO DETERMINE IF THEY LEAD TO SAFETY AND HEALTH PROGRAM OBJECTIVES AND GOAL.

Time and resources can be squandered when safety and health program activities do not achieve the desired results. Therefore, each year activities should be planned to achieve objectives that lead to the goal set for that year's safety and health program. The yearly evaluation assesses that achievement. (For information about setting a goal and objectives, see Pamphlet #2 in the Series, "Determining the Direction of Your Safety and Health Program: Establishing Policy, Goal and Objectives.")

Plan = Goal → Objectives → Procedures → Activities

To begin an evaluation you reverse the process, comparing program activities and the procedures that govern them to program objectives. Then evaluate the objectives against the goal of the program. If this analysis finds program efforts that are not effective, recommendations for changes in the program can be made for the next year.

Evaluation = Activities → Procedures → Objectives → Goal

### Activities and Procedures

Do the actual safety and health program activities and the procedures for organizing them bring the results expected when the safety and health program was written?

Evaluation = **Activities** → **Procedures** → Objectives → Goal

*Smaller worksites.* Even if yours is a smaller business, you should stop and think about the manner in which safety and health activities were carried out for the period evaluated and whether the results achieved were those expected at the outset.

*Larger worksites.* Large companies will have written procedures for the major activities of their safety and health program. They will also have written records of those activities as they were performed. Evaluating whether or not the written procedures were followed in the period evaluated or how well they were followed is an audit function of quality assurance or program evaluation.

*Sample questions.* The precise questions you should ask and get answered during the audit portion of the evaluation will depend upon the activity audited and the way it was planned to be accomplished. For example, if a certain person or group of people are to carry out the inspection program, were they the ones who actually did the inspections? (In many workplaces, inspections are conducted by the person with the most expertise and members of the site safety and health committee.) Was the "expert" on every inspection? Did the employee members always participate?

Other questions about inspections might include the following: Is there evidence that the "inspectors" went to every part of the worksite that was planned for each inspection? Did their reports indicate that they were finding the kinds of hazards that they were trained to recognize? Did hazard correction get appropriately assigned? Did the hazards get corrected in an appropriate and timely manner? Was the correction tracked to completion?

Similar questions should be asked and answered about each activity under the safety and health program. When a difference is found between the way the activity was planned and the way it is being carried out, assess which way best meets the objectives and goal. Then make sure that everyone follows that procedure.

### Objectives

The objectives connect the goal for the safety and health program to the program procedures and activities.

Evaluation = Activities → Procedures → **Objectives** → Goal

*Objectives that can be "audited."* Sometimes, when setting up new activities or improving existing ones, one or more of the objectives will be to complete these activities. For example, suppose the objective is "to complete one job safety analysis *each month*, with follow-up revision of safe work procedures and employee training in the *following month*." In this case, the objective describes the *frequency* of activities rather than the result. Because the question is simply whether the activities occurred, not what is the *quality* of the results, the evaluation would be more of an audit. Since the objective is being audited, look for evidence that job safety analyses were done each month. Is there evidence that revisions of procedures and training were also made each month following the job safety analysis that was done the month before? If the answer is "yes," and other program evaluation reveals no need to do something different, the frequency of these activities would become one of the ongoing procedures to be audited.

*Objectives that must be "evaluated."* Ordinarily, objectives should be aimed at the results desired from the program activities. For example, you might have an objective "to identify and assign all areas of safety and health responsibility that are not now clearly assigned, so that all safety and health responsibilities can be carried out successfully." There would then be a set of activities needed to accomplish that objective. The activities might include assigning a committee to implement the objective by listing all the safety and health responsibilities; reviewing assignment of those responsibilities; identifying missing assignments, duplicate assignments and unclear assignments; and recommending clear assignments.

In this case the evaluation would focus on whether the objective was accomplished. That is, were all areas of safety and health responsibilities that were not clearly defined actually identified and clarified? Perhaps these areas were identified and corrected, but the corrections have not been made because the Personnel Department still must rewrite seven major job descriptions to get the responsibilities correctly assigned. This, then, leads to a new objective, to get the seven job descriptions rewritten with clearer assignment of safety and health responsibilities. If all areas were identified and clarified, this objective would not lead to another objective.

If, to take another example, the objective is to increase employee safety and health awareness by involving the employees' families in a safety and health awareness program, the evaluation would begin with some sort of measurement of the difference between employee awareness before the family program and after it. One way to measure that difference might be to ask the employees who participated in the program to complete a questionnaire. Another would be to select employees randomly to ask if they thought their awareness was raised. Still another would be to ask supervisors if they saw a difference in employee awareness after the family program began. If, after using some type of meaningful measurement tool, the findings were that the program had increased awareness, the family program would be an ongoing activity subject to audit. If the findings were that no increase could be measured, a different family program could be tried to implement the same objective; or a different objective could be set to increase employee awareness.

### Goal

Evaluation = Activities → Procedures → Objectives → **Goal**

The goal provides the framework for the safety and health program. Objectives should be evaluated to be sure they are leading to the program goal. For example, suppose the goal was the reduction of employee exposure to hazards, and one of the objectives to achieve that goal was to hold safety meetings for all employees every month. Since this is an activity objective, it could be audited to make sure that safety meetings were held. But the next question is, "Did meeting this objective help meet the goal of the safety and health program?" In other words, did the meetings help employees understand the hazards to which they are exposed and result in plans to reduce that exposure?

Here are several ways to collect that information. If employees are asked to sign off on a sheet to show that they attended the training, they could also be asked to describe the hazards and potential hazards that exist in their work area, and how to control them better. If there was no sign-off sheet, interviews with some randomly selected employees could reveal their assessment of whether the meetings helped their understanding of the hazards and resulted in plans to control them. Interviews with supervisors would also show whether employees showed better understanding after training.

If the results indicate that not much was learned at these sessions, ask further questions to see what went wrong. In this manner, all the objectives should be checked to see if they helped achieve the goal. Stopping to evaluate programs against the goal of the safety and health program can avoid wasting time on an activity which is not really helping to make the worksite a safer, more healthful place to work.

### Evaluation Judgments

Gathering information about activities under the safety and health program is very important and the most time-consuming part of program evaluation. It is, however, the easiest to understand and to do. The hardest part is making judgments about program effectiveness. Without good judgments, the evaluation becomes a waste of time. With good judgments, it can be the most important part of the safety and health program.

Some sample sets of questions shown in the sections above can be used to guide the evaluator to judgments. There is also a sample section of evaluation instructions shown in Appendix 2 at the back of this pamphlet.

The evaluation should be viewed as determining the "bottom line" in the sense of the program goal "profit." The "profit" in program activity is any contribution to the achievement of your goal. All of the decisions that evaluators arrive at should be driven by the concept, "What does it do to contribute to the goal?"

To assist smaller businesses in making these judgments, there is a sample checklist, Appendix 1, starting on page 14 of this pamphlet that could be used to help decide what information should be gathered and what judgments need to be made.

Worksites should draw up their own specific procedures for gathering information and making judgments. They may add environmental, product safety, or security aspects to the system as well. The questions that need to be asked to guide the judgment should be based on individual site program activities and site objectives. (See the sample questions in Appendix 3 of this pamphlet.)

### Summary

Every year each employer should take some time to look at the activities of the safety and health program and evaluate their effectiveness in reaching the program goal. Activities should be audited against their procedures, the objectives set for the procedures, and the goal of the safety and health program. This process will give you good information about the quality of your safety and health program, help you to improve where needed, and to avoid wasting time and resources on activities that really do not help.

## HOW TO USE THE EVALUATION

The evaluation will only prove valuable if it results in improvements in meeting the safety and health goal. Some of the recommendations that result from the evaluation will be for one-time corrections. Many, however, will involve changing emphasis or trying new program activities. These recommendations should be incorporated into the objectives for the next year. Think about setting up one "permanent" objective "to audit the procedures set for safety and health program activities."

### Larger Worksites

The evaluation should result in a written report with written recommendations and documented follow-up to those recommendations. It may be useful to refer to past years' evaluations when preparing new ones or when planning new objectives. If you find that the same recommendations are being made year after year, the system of tracking recommendations to completion needs improvement.

### Smaller Worksites

At smaller sites where a written evaluation report may not be practical, make sure you set aside time to think about what you want to change. You have thought about what you did during the course of the year; you have talked to people; you have looked at your working conditions; and you have reviewed the documents that you do keep. Now decide what you want to do differently, and make sure that everyone understands what is expected from now on.

### Summary

Good quality control or evaluation results in ideas that you can use to get better quality in your safety and health program next year and beyond.

## STARTING OR OVERHAULING A SELF-EVALUATION SYSTEM

### Establish the Areas to Be Evaluated

To decide what needs to be evaluated, start by looking at the list of items in the section above called "What Should Be Evaluated." The sample checklist for small businesses (Appendix 1 of this pamphlet) can also serve as the core of the things you should want to evaluate. Small businesses may not need to add much to that group.

Larger businesses need more detail. For example, the checklist for small businesses shows only one indicator for routine inspections; i.e. "routine inspections are conducted regularly by adequately trained personnel." Larger worksites will want to spell out more detail, such as whether inspections follow the procedures set for them, whether they are finding the hazards which they should, whether the hazards are tracked to correction, or whether the same hazards keep reoccurring. These thoughts should also be part of the evaluation in a small business, but they would not necessarily need to be spelled out so clearly.

Be sure that all objectives set for your safety and health program are included in the evaluation.

### Select the Evaluator

The evaluator should be familiar with the activities used in the safety and health program but not be intimately involved in overseeing or implementing them. He/she should know safety and health program management.

### Use Self-Evaluation Tools

Make sure that whoever does the evaluation knows how to gather the information to be used in making the evaluation judgments.

Help is provided in Appendix 2, "Further Description of Safety and Health Program Assessment," found on page 22 of this pamphlet.

### Provide Instructions to Evaluate Activities Against Procedures, Objectives, and Goal

Make sure that whoever does the evaluation knows that the activities should be evaluated against procedures, objectives and the goal of the program. A good evaluation report format can help you accomplish that. (See the example of a section of a format for a narrative report in Appendix 3 of this pamphlet.)

Insist that the evaluator find the "bottom line," i.e., what has this activity done to contribute to the goal? An evaluation which provides no insight into the program should be rejected. In particular, do not accept a narrative which only describes the programs as they are designed. Insist that the hard questions about the effectiveness of the program be asked and answered.

### Use the Evaluation to Determine New Objectives

When the evaluation is complete, you should have a list of new items that you wish to make objectives. Check last year's objectives to see which of them need to be continued and which are either completed or do not help reach your goal. More information about setting objectives can be found in Pamphlet #2 in the series, "Determining the Direction of Your Safety and Health Program: Establishing Policy, Goal and Objectives."

### SUMMARY

This pamphlet has explained what a safety and health audit and evaluation is and how to accomplish this task. The body of the pamphlet describes what should be evaluated, who should do the evaluation and with what tools, how the evaluation should be done, how to use results of the evaluation, and how to begin or to overhaul the evaluation system you are using.

## APPENDICES

Appendix 1 provides an example of a self-evaluation checklist for small businesses that may be used to perform a comprehensive evaluation of the safety and health program.

Appendix 2 provides a step by step guide to using the three evaluation tools: documentation review, employee interviews, and hazardous conditions review and analysis—to assess each element and sub-element of your safety and health program.

Appendix 3 provides a sample evaluation instruction sheet that can be used to evaluate routine inspections when such inspections are part of your safety and health program.

Using the information in Chapter 13, and in Appendices 1, 2, and 3 to evaluate your safety and health program annually, you will be able to see clearly the bottom line in safety and health matters just as you now can scrutinize the bottom line in financial matters. In smaller companies, plant management will have the information they need to provide quality safety and health protection to all employees. In larger companies, by reporting this information to Boards of Directors, management at the highest levels can make knowledgeable and effective decisions to improve workers' safety and health. Corporate leaders, closer to the employees than OSHA, can lead the way to ensuring safe and healthful working conditions for America's working men and women.

Appendix 4 provides a list of 12 pamphlets published, currently in draft form, in the Safety and Health Program Assistance Series.

Appendix 5 is a listing of Safety Training Materials vendors. This list was provided courtesy of Occupational Health and Safety Magazine.

# APPENDIX 1

# EXAMPLE SELF-EVALUATION CHECKLIST FOR SMALL BUSINESSES

*Instructions:* Fill out the last three columns after judging the *effectiveness* of each part of your safety and health program. If the part of your program indicated in the first column is *fully* effective, enter "yes" in the second column. If not, enter "no," and explain the deficiencies in the third column marked "comments." If specific improvements are planned, enter the date they are *completed* in the last column.

**ELEMENT I. Management Commitment and Employee Involvement**

| Indicators | Yes/No | Comments |
|---|---|---|
| Date | | |
| Completed | | |
| Safety and Health responsibility is clearly assigned. | | |
| Adequate authority and resources given to those with responsibility. | | |
| Managers, supervisors and employees are held accountable. | | |
| Program operations are reviewed at least annually to evaluate success in meeting goals and objectives and prepare new objectives. | | |

**ELEMENT I. Management Commitment and Employee Involvement (*Continued*)**

| Indicators | Yes/No | Comments |
|---|---|---|
| Date | | |
| Completed | | |
| Policies and objectives established and communicated to all employees. | | |
| Top management is visibly involved in safety and health. | | |
| Employees are involved in problem finding and solving. | | |

**ELEMENT II. Worksite Analysis**

| Indicators | Yes/No | Comments |
|---|---|---|
| Date | | |
| Completed | | |
| A baseline comprehensive survey of hazards has been done or updated within the last three years. | | |
| Change analysis is done for every change of facility, equipment, process or materials. | | |
| Job hazard analysis is done on an ongoing basis. | | |
| Self-inspections are conducted regularly by adequately trained persons. | | |
| Routine inspections are conducted regularly by adequately trained personnel. | | |

**ELEMENT II. Worksite Analysis (*Continued*)**

| Indicators | Yes/No | Comments |
|---|---|---|
| Date | | |
| Completed | | |
| Employees know how and whom to notify about hazards, without fear of reprisal, and receive timely and appropriate responses. | | |
| Accidents and "near miss" incidents are investigated to identify all contributing causes and to prevent future occurrences. | | |
| Injury and illness experience are reviewed over a period of time long enough to establish patterns which may have common causes. | | |

**ELEMENT III. Hazard Prevention and Control**

| Indicators | Yes/No | Comments |
|---|---|---|
| Date | | |
| Completed | | |
| All identified hazards are prevented or controlled in the best feasible manner. | | |
| Safe work procedures based on job hazard analyses have been established. | | |
| Supervision reinforces safe work through positive feedback and training. | | |
| Enforcement of safe work procedures and safety and health rules is accomplished fairly and efficiently through a disciplinary system which all employees understand. | | |
| New or repeat hazards are identified and corrected in a timely manner. | | |

**ELEMENT III. Hazard Prevention and Control (*Continued*)**

| Indicators | Yes/No | Comments |
|---|---|---|
| Date | | |
| Completed | | |
| The facility and equipment are maintained regularly to prevent hazardous breakdowns. | | |
| Arrangements with Occupational Health Specialists provide medical services, including assistance in health problem identification. | | |
| First aid and CPR-trained employees are available on every shift. | | |
| Emergencies of all types have been prepared for. | | |
| Exits, evacuation routes and emergency telephone numbers are prominently displayed. | | |

**ELEMENT IV. Safety and Health Training**

| Indicators | Yes/No | Comments |
|---|---|---|
| Date | | |
| Completed | | |
| Employees can explain how and why they do the job safely and healthfully. | | |
| Employees can explain why PPE is used, how to use and maintain it and what the limits of its protection are. | | |
| Supervisors can explain safety rules and procedures for hazard control, how they teach this to employees and how they enforce this. | | |
| Managers can explain their safety and health responsibilities. | | |

**Special Objectives (List and evaluate individually)**

| Indicators | Yes/No | Comments |
|---|---|---|
| Date | | |
| Completed | | |

## APPENDIX 2

# FURTHER DESCRIPTION OF A SAFETY AND HEALTH PROGRAM ASSESSMENT

### I. INTRODUCTION

This description follows the outline on pages 2–4 of this pamphlet. It explains each of the assessment tools and how to use them to assess each element and sub-element of a safety and health program.

There are three basic methods for assessing safety and health program effectiveness. These are:

1. Checking documentation of activity;
2. Interviewing employees at all levels for knowledge, awareness and perceptions; and,
3. Reviewing site conditions and, where hazards are found, tracing the root causes of those hazards.

Some elements of program management better lend themselves to one method of assessment than do others. Some elements should be assessed using two or all three of the methods.

## Documentation

Checking documentation is a standard audit technique. It is particularly useful for understanding whether or not the tracking of hazards to correction is effective, but it can also be used to ascertain the quality of certain activities such as self-inspections or routine hazard analysis.

Where documentation is done, inspection records can tell the assessor/auditor whether or not serious hazards are being found or whether the same hazards are found over and over again. In the first case, if serious hazards are not being found and yet accidents keep occurring, there may be a need to train those who inspect to look for different hazards than they do now.

In the second case, if the same hazards are being found over and over again, the problem may be more complicated. Perhaps the hazards are not being corrected. That would be a tracking problem or a problem with accountability for hazard correction.

If the same hazards occur over and over again after they are corrected, someone is not taking responsibility for keeping those hazards under control. Either the responsibility is not clear or those who are responsible are not being held accountable.

## Employee Interviews

Talking to randomly selected employees at all levels will provide a good indication of the quality of employee training and of employee perceptions of how well the program is running. If the training you are providing is effective, the employees you interview will be able to tell you about the hazards they work with and how they help to protect themselves and others by keeping those hazards controlled. Every employee should also be able to say precisely what he or she is expected to do, and where to go if an emergency occurs.

Employee interviews can tell you about other aspects of your program as well. The employee's perception of how easy it is to report a hazard (and get a timely and appropriate response) will tell you a lot about how well your system for employee reports of hazards is working. Employee perception of your system for enforcing safety and health rules and safe work practices can also be enlightening. If they perceive inconsistency or confusion, you will know that the system needs improvement.

Employee interviews are not limited to hourly employees. A lot may be learned from talking with first-line supervisors. It is also helpful to find out from some line managers what their perceptions of their safety and health responsibilities are.

## Site Conditions and Root Causes of Hazards

Looking at the conditions of the workplace will reveal the hazards that exist there. It can also provide a lot of information about the breakdown of management systems meant to prevent or control those hazards.

Looking at conditions and practices is well-understood as a technique of assessing the effectiveness of safety and health programs. For example, if, in areas where personal protective equipment is required, you see large and understandable signs designating areas where PPE is required and employees in those areas wearing equipment properly (with no exceptions), you have seen a valuable clue as to how well the PPE program is working.

In addition to noting good conditions and practices, you can also obtain good information about management from the hazards that are seen through *root analysis.* This approach to hazards is much like the most sophisticated accident investigation techniques in which many contributing factors are located and corrected or controlled.

For example, if, during a review of conditions, you find a machine being operated without a guard on a pinch point, you should not stop with just getting the guard put on. Asking a few questions will tell you a lot about one or more widely varied management systems. If the operator says he did not know the guard was supposed to be on the machine, that leads to a series of questions about the existence of safe work procedures and/or training.

If he says that the guard slows him down and that the supervisor knows he takes it off, that leads to a lot of questions about enforcement of rules, accountability and the clarity of the worksite objective of putting safety and health first.

If, on the other hand, the need for the guard is noticed by your insurance inspector or an OSHA inspector—or you do not notice it until someone is hurt, the questions to be asked should start with whether or not a comprehensive survey of the worksite has been done by someone with enough expertise to recognize all the potential and existing hazards.

Analyzing the root causes of hazards is really a technique that should not wait for formal assessments of your program. This technique should be in use every day. However, when used as part of a formal assessment, it is very helpful.

To assess each part of your worksite's safety and health program, you should use one or more of these methods, as appropriate. The remainder of this appendix will identify the sub-elements found in each element of a quality safety and health program and identify methods which may be useful in assessing the sub-element.

## II. ASSESSING THE KEY SUB-ELEMENTS OF MANAGEMENT COMMITMENT

### Worksite Policy on Safe and Healthful Working Conditions

#### *a. Documentation*

If there is a written policy, does it make clear the priority of worker safety and health in relation to other organizational values, such as production?

### *b. Interviews*

1. When asked, can employees at all levels give you a fair idea of what your worksite policy on worker safety and health is?
2. If the policy is written, can hourly employees tell you where they have seen it?
3. Can employees at all levels explain the priority of worker safety and health in relation to other organizational values, as the policy intends?

### *c. Site Conditions and Root Causes of Hazards*

Have injuries occurred because employees at any level did not understand the importance of safety precautions in relation to other organizational values, such as production?

## Goal and Objectives for Worker Safety and Health

### *a. Documentation*

1. If there is a written goal for your safety and health program, is it updated annually?
2. If there are written objectives, such as an annual plan to reach that goal, are they clearly stated?
3. If managers and supervisors have written objectives, do they include objectives for the safety and health program?

### *b. Interviews*

1. Do managerial and supervisory employees have a clear idea of their objectives for worker safety and health?
2. Do hourly employees understand the current objectives of the safety and health program?

### *c. Site Conditions and Root Causes of Hazards*

(Only helpful in a general sense.)

## Visible Top Management Involvement

### *a. Documentation*

Are there one or more written programs which provide for the involvement of top-level management in safety and health activities? For example, top management can receive and sign off on inspection reports received either after each inspection or in a quarterly summary. These reports can then be posted for employees to see. Top management can provide "open door" times each week or each month for employees to come in to discuss safety and health concerns. Top management can reward the best safety suggestions each month or at other specified intervals.

***b. Interviews***

1. Can hourly employees describe how management officials are involved in safety and health activities?
2. Do hourly employees perceive that managers and supervisors follow safety and health rules and work practices such as wearing appropriate personal protective equipment?

***c. Site Conditions and Root Causes of Hazards***

When employees are found not wearing required personal protective equipment or not following safe work practices, have any of them said that managers or supervisors also did not follow these rules?

### Employee Involvement

***a. Documentation***

1. Are there one or more written programs that provide for employee involvement in decisions which affect their safety and health?
2. Is there documentation of those activities (such as employee inspection reports, minutes of joint employee-management or employee committee meetings)?
3. Is there written documentation of any management response to employee safety and health program activities?
4. Does the documentation indicate real substance to employee activities?
5. Are there written guarantees of protection for employees from possible harassment resulting from safety and health program activities?

***b. Interviews***

1. Are employees aware of the way(s) in which they can be involved in decisions that affect their safety and health?
2. Do employees appear to take pride in the achievements of the worksite safety and health program?
3. Are employees comfortable answering questions about safety and health programs and conditions at the site?
4. Do employees feel that they have the support of management for their safety and health activities?

***c. Site Conditions and Root Causes of Hazards***

(Not applicable.)

## Assignment of Responsibility So That Everyone Knows What Performance Is Expected

### *a. Documentation*

Are responsibilities written out so that they can be clearly understood?

### *b. Interviews*

Do employees understand their own responsibilities and those of others?

### *c. Site Conditions and Root Causes of Hazards*

1. Was the hazard caused in part because no one was assigned the responsibility to control or prevent it?
2. Was the hazard allowed to exist in part because someone in management did not have the clear responsibility to hold a lower-level manager or supervisor accountable for carrying out assigned responsibilities?

## Adequate Authority and Resources So That Responsibilities Can Be Met

### *a. Documentation.*

(Only generally applicable.)

### *b. Interviews*

1. Do safety staff members or any other personnel have responsibilities to ensure safe operation of production equipment without the authority to shut down that equipment or order maintenance or parts?
2. Do employees talk about not being able to get safety or health improvements because of cost?
3. Do employees mention the need for more safety or health personnel or expert consultants?

### *c. Site Conditions and Root Causes of Hazards*

1. Do recognized hazards go uncorrected because of lack of authority or resources to correct them?
2. Are there unrecognized hazards because better expertise is needed to diagnose hazards?

## Holding Managers, Supervisors and Hourly Employees Accountable for Meeting Their Safety and Health Responsibilities

### *a. Documentation*

1. Do all line manager and supervisory performance evaluations include specific safety and health protection criteria?
2. Is there documented evidence of employees at all levels being held accountable for safety and health responsibilities, including safe work practices, either through performance evaluations impacting pay and/or promotions or disciplinary actions?

***b. Interviews***

1. When you ask employees what happens to people who break safety and health rules or safe work practices, do they indicate that such employees are held accountable in a clear and consistent manner?
2. Do hourly employees indicate that supervisors and managers really care about meeting safety and health responsibilities?
3. When asked about what happens when rules are broken, do hourly employees complain that supervisors and managers do not follow rules and are never disciplined for those infractions?

***c. Site Conditions and Root Causes of Hazards***

1. Are hazards occurring because employees, supervisors and/or managers are not being held accountable for their safety and health responsibilities?
2. Are identified hazards not being corrected because those assigned the responsibility are not being held responsible?

### Evaluate Program Operations

***a. Documentation***

Is there a written evaluation of each major part of the program, as listed in the OSHA Safety and Health Program Management Guidelines (54 FR 3908, January 26, 1989) which lists what is being done, evaluates how effective it is against the goal and objectives, and recommends changes where needed to make the program more effective or to try other programs instead?

***b. Interviews***

Can employees, supervisors and/or managers tell you how the whole program is evaluated and revised each year?

***c. Site Conditions and Root Causes of Hazards***

(Only generally applicable.)

## III. ASSESSMENT OF KEY SUB-ELEMENTS OF WORKSITE ANALYSIS

### Comprehensive Surveys, Change Analysis, Routine Hazard Analysis

***a. Documentation***

1. Are there documents which provide comprehensive analysis of all potential safety and health hazards of the worksite?
2. Are there documents which provide analysis of potential safety and/or health hazards for each new facility, equipment, material or process and the means for elimination or control of any hazards identified?
3. Is the step-by-step analysis of the hazards in each part of each job documented so that you can clearly see the analysis that resulted in decisions on safe work procedures?

4. If there are complicated processes with a potential for catastrophic impact from an accident but low probability of an accident, such as in nuclear power or chemical production, are there documents analyzing the potential hazards in each part of the processes, and the means to prevent or control them?
5. If there are processes with a potential for catastrophic impact from an accident but low probability of an accident, have analyses of the processes such as "fault tree" or "what if?," been documented to ensure enough back-up systems for worker protection in the event of more than one control failure?

### b. Interviews

1. Do employees complain that new facilities, equipment, materials or processes are hazardous?
2. Do any employees say that they have been involved in job safety analysis or process review and are satisfied with the results?
3. Does the safety and health staff indicate ignorance of existing or potential hazards at the worksite?
4. Does the occupational nurse/doctor or health care provider understand the potential occupational diseases and health effects in this worksite?

### c. Site Conditions and Root Causes of Hazards

1. Have hazards appeared where no one in management realized there was potential for their development?
2. Where workers have followed job procedures faithfully, have accidents or "near misses" occurred because of hidden hazards?
3. Have hazards been discovered in the design of new facilities, equipment, materials and processes after use was begun?
4. Have accidents or "near misses" occurred when two or more failures in the hazard control system surprised everyone by occurring at the same time?

## Regular Site Safety and Health Inspections

### a. Documentation

1. If inspection reports are written, do they show that inspections are done on a regular basis?
2. Do the hazards found indicate good ability to recognize hazards which are typical for this industry?
3. Are hazards found during inspections tracked to complete correction?
4. What is the relationship between the hazards found in inspections and those found to be at least in part responsible for injuries or illness?

#### *b. Interviews*

Do employees indicate that they see inspections being done that seem thorough?

#### *c. Site Conditions and Root Causes of Hazards*

Are the hazards discovered during accident investigations ones that should have been recognized and corrected by the inspection process?

### Employee Reports of Hazards

#### *a. Documentation*

1. Is the system for written reports being used frequently?
2. Are valid hazards that have been reported by employees tracked to complete correction?
3. Are the responses timely and adequate?

#### *b. Interviews*

1. Do employees know whom to contact and what to do if they see something which they believe to be hazardous to themselves or coworkers?
2. Do employees feel that responses to their reports of hazards are timely and adequate?
3. Do employees mention that sometimes when they report a hazard they never hear anything further about it?
4. Do any employees mention that they or other employees are being harassed officially or otherwise for reporting hazards?

#### *c. Site Conditions and Root Causes of Hazards*

1. When hazards are found, is it ever the case that employees say they have been complaining about them for a long time but no one has done anything to achieve correction?
2. Are hazards ever found where you would think employees would have been sure to recognize them but never reported them?

### Accident and "Near Miss" Investigations

#### *a. Documentation*

1. Do accident investigation reports show a thorough analysis of the cause(s) of accidents, rather than quickly blaming the injured employee?
2. Are "near misses" (property damage or "close calls") investigated using the same techniques as accident investigations?
3. Are hazards which are identified as contributing to accidents or "near misses" tracked to correction?

#### b. Interviews

1. Do employees understand and accept the results of accident and "near miss" investigations?
2. Do employees mention a tendency on management's part to blame the injured employee?
3. Do employees feel that all hazards contributing to accidents are corrected or controlled?

#### c. Site Conditions and Root Causes of Hazards

Are accidents sometimes caused at least in part by factors which might have contributed to "near misses" which were not investigated or to accidents which were too quickly investigated?

### Injury and Illness Pattern Analysis

#### a. Documentation

1. In addition to the required OSHA log, are careful records kept of first aid injuries and/or illnesses that might not immediately appear to be work-related?
2. Is there any periodic, written analysis of the patterns of "near misses," injuries and/or illness over time to see if there are previously unrecognized connections between them which indicate unrecognized hazards needing correction or control?
3. Looking at the OSHA 200 log and, where applicable, first aid logs, are there patterns of illness or injury that should have been analyzed for previously undetected hazards?
4. If there is an occupational nurse/doctor on the worksite or if employees are encouraged to take ordinary illness to a nearby health care provider, are the lists of those visits analyzed to see if there are any clusters of illness which might be work-related?

#### b. Interviews

Have employees mentioned illnesses or injuries that seem related to them but which have not been analyzed for previously undetected hazards?

#### c. Site Conditions and Root Causes of Hazards

(Not generally applicable.)

# IV. ASSESSMENT OF KEY SUB-ELEMENTS OF HAZARD PREVENTION AND CONTROL

## Appropriate Use of Engineering Controls, Work Practices, Personal Protective Equipment and Administrative Controls

### a. *Documentation*

1. If there are documented comprehensive surveys, are they accompanied by a plan for systematic prevention or control?
2. If there is a written plan, does it show that the best method of hazard protection was chosen?
3. Are there written safe work procedures?
4. If respirators are used, is there a written respirator program?

### b. *Interviews*

1. Do employees say that they have been trained in and have ready access to reliable safe work procedures?
2. Do employees mention that they have difficulty accomplishing their work because of unwieldy controls meant to protect them?
3. Do employees ever mention personal protective equipment (PPE), work practices or engineering controls as interfering with their ability to work safely?
4. Do employees who use PPE understand why they use it and how to maintain it?
5. Do employees who use PPE indicate that the rules for PPE use are consistently and fairly enforced?
6. Do employees indicate that safe work procedures are fairly and consistently enforced?

### c. *Site Conditions and Root Causes of Hazards*

1. Do you ever find that controls meant to protect workers are actually putting them at risk or not providing enough protection?
2. Are employees involved in unsafe practices or creating unsafe conditions because rules and work practices are not fairly and consistently enforced?
3. Are employees in areas designated for PPE wearing it properly and without exception?
4. Are hazards which could feasibly be controlled by better design being inadequately controlled by other means?

## Provide Facility and Equipment Preventive Maintenance

### *a. Documentation*

1. Is there a preventive maintenance schedule that provides for timely maintenance of the facilities and equipment?
2. Is there a written or computerized record of when maintenance was performed which shows that the schedule has been followed?
3. Do maintenance request records show a pattern of certain facilities or equipment needing repair or breaking down before maintenance was scheduled or before it was actually done?
4. Do any accident/incident investigations list facility or equipment breakdown as a major cause?

### *b. Interviews*

1. Do employees mention difficulty with improperly functioning equipment or facilities in poor repair?
2. Do maintenance employees feel that the preventative maintenance system is working well?
3. Do employees feel that hazard controls needing maintenance are properly cared for?

### *c. Site Conditions and Root Causes of Hazards*

1. Is poor maintenance a frequent source of hazards?
2. Are hazard controls in good working order?
3. Does equipment appear to be in good working order?

## Plan and Prepare for Emergencies

### *a. Documentation*

Are there written procedures that clearly cover every likely emergency with clear evacuation routes, assembly points and emergency telephone numbers?

### *b. Interviews*

Can employees tell you exactly what they are supposed to do in case of any kind of likely emergency that you ask them about?

### *c. Site Conditions and Root Causes of Hazards*

1. Have hazards occurred in actual or simulated emergencies due to confusion about what to do?
2. In larger worksites, are emergency evacuation routes clearly marked?
3. Are emergency telephone numbers and fire alarms easy to find?

### Establish a Medical Program

#### *a. Documentation*

Are good, clear records kept of medical testing and assistance?

#### *b. Interviews*

1. Do employees say that test results were explained to them?
2. Do employees feel that more first aid or CPR-trained personnel should be available?
3. Are employees satisfied with the medical arrangements at or for the site?
4. Does the occupational health care provider understand the potential hazards of the worksite so that occupational illness symptoms can be recognized?

#### *c. Site Conditions and Root Causes of Hazards*

1. Have further injuries or worsening of injuries occurred because proper medical (including trained first aid and CPR) assistance was not readily available?
2. Have occupational illnesses gone undetected because no one with occupational health specialty training reviewed employee symptoms as part of the medical program?

## V. ASSESSMENT OF KEY SUB-ELEMENTS OF SAFETY AND HEALTH TRAINING

### Ensuring That All Employees Understand Hazards

#### *a. Documentation*

1. Does the written training program include complete training for every employee in every potential hazard to which each employee may be exposed and in emergency procedures?
2. Do training records show that every employee received the planned training?
3. Do the written evaluations of training indicate that the training was successful and that the trained employees learned what was intended?

#### *b. Interviews*

1. Can employees tell you what hazards they are exposed to, why those hazards are a threat, and how they can help protect themselves and others from those hazards?
2. If PPE is used, can employees explain why they use it and how to use and maintain it?

3. Can employees tell you precisely what they are supposed to do and where they are supposed to go in every kind of emergency likely to occur at your worksite?
4. Do employees feel that health and safety training is adequate?

#### c. Site Conditions and Root Causes of Hazards

1. Have employees been hurt or made ill by hazards of which they were completely unaware, or did not understand the danger, or from which they did not know how to protect themselves?
2. Have employees or rescue workers ever been endangered by employee lack of knowledge about what to do or where to go in any given type of emergency situation?
3. Are there hazards in the workplace that exist, at least in part, because one or more employees was not adequately trained in hazard control?
4. Are there any instances of employees not wearing required PPE properly because they have not been properly trained?

### Ensuring That Supervisors Understand Their Responsibilities

#### a. Documentation

Do training records indicate that all supervisors have been trained in their responsibilities to analyze work under their supervision for unrecognized hazards, to maintain physical protection and to reinforce employee training through performance feedback and, where necessary, enforcement of safe work practices and safety and health rules?

#### b. Interviews

1. Are supervisors aware of their responsibilities?
2. Do employees say that supervisors are carrying out these duties?

#### c. Site Conditions and Root Causes

Has supervisor lack of understanding of safety and health responsibilities played a part in creating hazardous activities or conditions?

### Ensuring That Managers Understand Their Safety and Health Responsibilities

#### a. Documentation

Do training plans and training records indicate that training on safety and health responsibilities is part of the training that managers receive and that all line managers have received it?

### *b. Interviews*
Do employees indicate that managers know and carry out their safety and health responsibilities?

### *c. Site Conditions and Root Causes of Hazards*
Has managerial lack of understanding of safety and health responsibilities played a part in creating hazardous activities or conditions?

# APPENDIX 3

# SAMPLE FROM AN EVALUATION INSTRUCTION*

Answer the questions/instructions for each section in narrative form with complete sentences. You may use one sentence to answer more than one question if appropriate. Your answers should reflect completed judgments. Avoid use of terms that reflect incomplete judgments such as "appears", "apparently" and "seems."

### Section 2. Routine Inspections

1. List any objectives set which involve routine inspections.
2. Are inspections following the procedures set for them? If not, describe how they are different and how frequently this difference occurs.
3. If procedures are not being followed, discuss the reasons and whether the alternative activity meets the objective set for inspections.
4. If any specific objectives were set for inspections, were they met effectively? Describe.
5. Has inspection activity contributed to identifying and controlling hazards and potential hazards? Describe.
6. If not otherwise covered above, please answer the following and provide examples:
   a. Are inspections revealing hazards slipping out of their controls, such as guards removed from machines, housekeeping problems or work

*The questions asked in this sample instruction are similar to those that should be asked for every element in your safety and health program if you are to achieve a thorough and meaningful evaluation.

procedures not being followed? If so, be sure to deal with this topic in the evaluation of hazard prevention and control.

b. Are inspections revealing new hazards that had not been identified before? If so, be sure to deal with this topic under the evaluation of comprehensive surveys, change analysis or routine hazard analysis.**

c. Are inspections revealing repeated instances of the same problem? If so, are they the same problem uncorrected or the same problem recurring? If the former, be sure to deal with this problem under the evaluation of tracking of hazard correction or accountability, or both. If the latter, be sure to deal with the problem under hazard prevention and control.***

7. List any recommendations for changes in routine inspection activity needed for next year. Make sure that the recommendations logically reflect the conclusions made above.

**For information about comprehensive surveys, change analysis, and routine hazard analysis, see Pamphlet 8, "Establishing Complete Hazard Inventories," of the series.

***For information about implementing programs for hazard prevention and control, see Pamphlet 10 of the series.

## APPENDIX 4

# LIST OF PAMPHLETS IN THIS SERIES[a]

| | |
|---|---|
| Pamphlet 1: | Managing Worker Safety and Health Protection: Introducing OSHA's Safety and Health Program Management Guidelines |
| Pamphlet 2: | Determining the Direction of Your Program: Establishing Policy, Goal, and Objectives |
| Pamphlet 3: | Visible Top Management Leadership: Showing Your Management Commitment |
| Pamphlet 4: | Employee Involvement |
| Pamphlet 5: | Assigning Safety and Health Responsibilities |
| Pamphlet 6: | Developing Accountability for Safety and Health |
| Pamphlet 7: | Safety and Health Program Quality Assurance: Determining How Well Your Safety and Health Program Works |
| Pamphlet 8: | Establishing Complete Hazard Inventories |
| Pamphlet 9: | Catching the Hazards That "Escape" Controls |
| Pamphlet 10: | Establishing Hazard Prevention and Control Programs |
| Pamphlet 11: | The Occupational Health Delivery System: A Medical Program That Is Right for Your Worksite |
| Pamphlet 12: | Safety and Health Training |

[a] Occupational Safety and Health Administration, Office of Consultation Programs, Safety and Health Assistance Series (September 1991 Draft).

# APPENDIX 5

Information provided courtesy of *Occupational Health & Safety* magazine.

# SAFETY TRAINING MATERIALS

The following numbers indicate specific topics:

| | | | |
|---|---|---|---|
| 1 | = AIDS Education | 10 | = Fire Fighting |
| 2 | = Audiovisual/Films/Filmstrips/ Slides | 11 | = Fire Prevention |
| | | 12 | = Hearing Analysis & Protection |
| 3 | = CPR | 13 | = Laboratory Safety |
| 4 | = Cartoons | 14 | = Lifting Techniques |
| 5 | = Computer Programs | 15 | = Machine Operator |
| 5a | = Confined Spaces | 16 | = Maintenance Safety |
| 6 | = Construction Safety | 17 | = Material Handling |
| 7 | = Driver Safety | 18 | = Noise Control |
| 8 | = Electrical Safety | 19 | = Office Safety |
| 9 | = Environmental Protection | 20 | = Rescue & Resuscitation |
| 9a | = Ergonomics | 21 | = Right-to-Know Compliance |

**Action Training Systems**
P.O. Box 5782
Vancouver, Wash. 98668
206-574-6471 or
800-262-9737
(7, 10, 17)

**Agress, M. Clifford, PE**
891 Fulton St.
Valley Stream, N.Y. 11580
516-825-8955
(8)

**AMBU**
611 N. Hammonds Ferry Road
Linthicum, Md. 21090-1356
410-636-1144 or
800-262-8462
(3, 20)

**American Society of Safety Engineers**
1800 E. Oakton St.
Des Plaines, Ill. 60018-2187
708-692-4121
(2)

**Armstrong Medical Industries**
575 Knightsbridge Pkwy.
P.O. Box 700
Lincolnshire, Ill. 60069-0700
708-913-0101 or
800-323-4220
(2, 3, 20)

**Associated Builders & Contractors**
729 15th St. NW
Washington, D.C. 20006
202-637-8800
(2, 6)

**Associated General Contractors of America**
1947 E St. NW
Washington, D.C. 20006
202-393-2040
(2, 6)

**Auburn Engineering Inc.**
123 N. College St.
Auburn, AL 36830
205-826-8600
(9a)

**Aurora Pictures**
2525 E. Franklin Ave.
Minneapolis, Minn. 55406
612-338-2825 or
800-346-9487
(2, 6, 7, 9a, 11, 12, 16, 18, 21)

**AXIA Services**
45 Glastonbury Blvd.
Glastonbury, Conn. 06033
203-659-6700 or
800-332-2942
(6, 7, 10)

**Bilsom International**
109 Carpenter Drive
Sterling, Va. 22170
703-834-1070 or
800-733-1177
(12, 18)

**BNA Communications**
9439 Key West Ave.
Rockville, Md. 20850
301-948-0540 or
800-233-6067
(2, 8, 9, 12, 13, 14, 15, 16, 17, 18, 19, 20, 21)

**Bongarde Communications**
P.O. Box 428
Oroville, Wash. 98844-0428
800-667-9300
(4, 7, 8, 11, 12, 14, 15, 16, 17, 18, 19, 20, 21)

**Broner Glove & Safety Co.**
359 Robbins Drive
Troy, Mich. 48083
800-521-1318
(8, 12, 14, 15, 21)

**Brown, Don Productions**
1429 W. Collins Ave.
Orange, Calif. 92667
714-771-9711 or
800-359-7910
(6, 7, 8, 9, 9a, 10, 11, 12, 14)

**Bureau of Business Practice**
24 Rope Ferry Rd.
Waterford, Conn. 06386
203-442-4365 or
800-243-0876
(7, 8, 16, 17, 19)

**Business & Legal Reports**
39 Academy St.
Madison, Conn. 06443
203-245-7448 or
800-727-5257
(2, 4, 7, 8, 12, 13, 14, 17, 18, 21)

**Cabot Safety Corp.**
90 Mechanic St.
Southbridge, Mass. 01550
508-764-5500 or
800-225-9038
(2)

**Capital Communications**
3807 Dickerson Road
Nashville, Tenn. 37207
615-868-2040 or
800-822-5678
(8, 12, 14, 15, 16, 17, 18, 19, 20)

**Cavanaugh Consulting Group**
421 E. Cook Rd. #200
Fort Wayne, Ind. 46825-3650
219-489-2963
(1, 5a, 6, 8, 9, 9a, 13, 14, 15, 16, 17, 18, 21)

**Center for Hazardous Materials Research (CHMR)**
320 William Pitt Way
Pittsburgh, Pa. 15238
412-826-5320
(9)

**Charlock Technologies & Associates**
P.O. Box 845
464-425 Church St.
Janesville, Calif. 96114
916-253-2547
(9)

**Coastal Video Communications**
3083 Brickhouse Court
Virginia Beach, Va. 23452
804-498-9014 or
800-767-7703
(7, 8, 9, 12, 13, 14, 17, 21)

**Comco Environmental & Safety Services**
17120 Clark Ave.
Bellflower, Calif. 90706-5730
310-804-0305 or
800-327-5570
(9)

**Comprehensive Loss Management (CLMI)**
15800 32nd Ave. N. #106
Minneapolis, Minn. 55447
612-551-1022 or
800-533-2767
(5a, 9a, 14, 19, 21)

**Comprehensive Safety Compliance**
295 William Pitt Way
Pittsburgh, Pa. 15238
412-826-5480
(6, 13, 14, 19, 21)

**Compressed Gas Associates**
1725 Jefferson David Hwy. #1004
Arlington, Va. 22202
703-979-0900
(2, 7, 10)

**Computer Information Applications**
275 Broadhollow Rd. #302
Melville, N.Y. 11747
516-756-2204
(2, 4, 5, 6)

**Conney Safety Products**
3202 Latham Drive
Madison, Wis. 53744
800-356-9100
(14)

**Crane Inspection & Certification Bureau**
5874 S. Semoran Blvd.
Orlando, Fla. 32822
407-277-0884 or
800-327-1386
(6)

**Crown Consulting Firm**
827 S. Union #110
Warsaw, Ind. 46580
219-268-1100
(20, 21)

**DBA**
111 E. Ave. Suite 321
Norwalk, Conn. 06851
203-853-1917 or
800-528-2810
(14, 17)

**Dependable Fire Equipment Service**
116 Telford Park
Telford, Pa. 18969
215-721-1633
(11)

**Detail Associates**
300 Grand Ave.
Englewood, N.J. 07031
201-569-6708
(9, 19, 21)

**Doron Precision Systems**
174 Court St.
P.O. Box 400
Binghamton, N.Y. 13902-0400
607-772-1610
(2, 7)

**Du Pont Safety & Environmental Resources**
Montgomery Bldg. 278
P.O. Box 80800
Wilmington, Del. 19880-0800
800-532-7233
(6, 7, 19)

**Educational Resources**
P.O. Box 1257
Lexington, S.C. 29071-1257
803-356-4880 or
800-333-8822
(7, 8, 11, 12, 14, 16, 17, 18, 19, 20, 21)

**Edwards, A Unit of General Signal**
195 Farmington Ave.
Farmington, Conn. 06032
203-678-0410
(2, 4)

**ELB & Associates**
605 Eastowne Drive
Chapel Hill, N.C. 27514
919-967-2228 or
800-334-5478
(1, 2, 8, 12, 13, 14, 21)

**EMED Co.**
P.O. Box 369
Buffalo, N.Y. 14240
716-626-1616 or
800-442-3633
(11)

**Emergency Film Group**
225 Water St.
Plymouth, Mass. 02360
508-746-0466 or
800-842-0999
(1, 2, 9, 21)

**Emergency Medical Update**
P.O. Box 11380
Bainbridge Island, Wash. 98110-5380
206-842-9775 or
800-327-3841
(1, 2, 3, 20)

**EMS Associates**
P.O. Box 380
Pawling, N.Y. 12564
914-878-3001 or
800-525-8901
(2, 20)

**Ensign Safety & Health Advisory**
300 W. Shaw Ave. #479
Clovis, Calif. 93612
209-299-0501
(14)

**Environmental & Occupational Health Sciences Institute**
45 Knightsbridge Road
Broodwood II
Piscataway, N.J. 08854-3923
908-463-5062
(9, 10, 11, 12, 13, 14, 15, 16, 17, 18, 19, 21)

**Environmental Technology Corp.**
11205 Alpharetta Hwy. B-4
Roswell, Ga. 30076
404-475-2055
(12)

**Ergodynamics Inc./Ergoworks Inc.**
2550 University Ave. #207N
St. Paul, Minn. 55114
612-646-0759 or
800-537-7475
(9a, 19)

**Ergodyne Corp.**
1410 Energy Park Drive, Suite 1
St. Paul, Minn. 55108
800-225-8238
(9a, 19)

**ERM Group**
855 Springdale Drive
Exton, Pa. 19341
215-524-3500 or
800-544-3117
(21)

**Excellence In Safety**
35 Sagamore Road
N. Weymouth, Mass. 02191
617-749-5449
(6, 15, 16, 19)

**Fire Prevention Through Films**
P.O. Box 11
44 Erie Ave.
Newton Highlands, Mass. 02161
617-965-4444
(10)

**Fire Safety Displays**
20422 Van Born Road
Dearborn Heights, Mich. 48125
313-274-7888
(11)

**Firecon**
P.O. Box 231
East Earl, Pa. 17519
717-354-2411 or
800-222-8841
(9, 10, 11)

**FLI Learning Systems**
P.O. Box 2233
Princeton, N.J. 08543-2233
609-466-9000
(7)

**Genium Publishing Corp.**
1145 Catalyn St.
Schenectady, N.Y. 12303-1836
518-377-8854 or
800-243-6486
(6, 13, 21)

**Government Institutes**
4 Research Place #200
Rockville, Md. 20850
301-921-2300
(9, 21)

**Halliburton NUS Environmental Corp.**
910 Clopper Road
Gaithersburg, Md. 20877
301-258-6000 or
800-368-2755

**Hand Tools Institute**
25 N. Broadway
Tarrytown, N.Y. 10591
914-332-0040
(9a)

**Hearing Conservation Products**
107 W. Broadway
Clarksville, Texas 75426
903-427-5621
(12, 18)

**Hogan Group**
123 Townsend St. #540
San Francisco, Calif. 94107
415-442-0210
(2, 19)

**Holt, RJ Sales & Marketing Associates**
1256 Kirk Road
Boothwyn, Pa. 19061
215-497-1766
(10, 18)

**Howell Training Co.**
5201 Langfield Road
Houston, Texas 77040
713-460-4460 or
800-527-1851
(5, 10, 21)

**Humantech Inc.**
173 Parkland Plaza, Suite D
Ann Arbor, Mich. 48103
313-663-6707
(9a)

**Hyster Co., North American Industrial Truck Division**
P.O. Box 847
Danville, Ill. 61834-0847
217-443-7000 or
800-221-1191
(7, 17)

**Industrial Hearing Testing**
55 Cedar St.
Worcester, Mass. 01609
508-752-4663
(12)

**Industrial Training Inc.**
5376 52nd St. SE
Grand Rapids, Mich. 49512
616-698-8688 or
800-253-4623
(1, 5a, 8, 9, 9a, 12, 13, 14, 15, 17, 21)

**Industrial Training Systems**
9 E. Stow Road
Marlton, N.J. 08053
609-983-7300 or
800-367-4871
(2, 5, 6, 7, 8, 9, 10, 11, 12, 13, 14, 15, 16, 17, 18, 19, 20, 21)

**Interactive Media Communications**
100 5th Ave.
Walthan, Mass. 02154
617-890-7707
(12, 13, 15, 17, 21)

**Interface Assoc.**
1000 Victors Way, Ste. 300
Ann Arbor, Mich. 48108
313-761-1698
(5a)

**ITC**
13515 Dulles Technology Drive
Herndon, Va. 22071-3416
800-638-3757
(2)

**IV Marketing**
P.O. Box 26869
Las Vegas, Nev. 89126-0869
702-253-9410 or
800-736-2754
(9)

**The Joyce Institute**
1313 Plaza 600 Bldg.
Seattle, Wash. 89101
206-441-6745
(9a)

**Kaufman, James A, & Associates**
101 Oak St.
Wellesley, Mass. 02181
617-237-1335 or
800-484-1005
(2, 8, 13, 16, 19)

**Keller, JJ & Associates**
3003 W. Breezewood Lane
P.O. Box 368
Neenah, Wis. 54957-0368
414-722-2848 or
800-558-5011
(2, 6, 7, 9, 13, 14, 16, 17, 21)

**Kemron Environmental Services**
7926 Jones Branch Drive #1100
McLean, Va. 22102
703-893-4106 or
800-777-1042
(21)

**Krames Communications**
1100 Grundy Lane, 2nd Floor
San Bruno, Calif. 94066-3030
415-742-0400
(1, 2, 6, 12, 13, 18, 19, 21)

**Lab Safety Supply Inc.**
P.O. Box 368
Janesville, Wis. 53547
800-356-0783 or
608-754-2345
(8, 9, 9a, 11, 12, 13, 14, 15, 16, 17, 18, 19, 21)

**Laerdal Medical Corp.**
1 Labriola Court
P.O. Box 190
Armonk, N.Y. 10504-0190
914-273-9404 or
800-431-1055
(3)

**Laser Institute of America**
12424 Research Pkwy. #130
Orlando, Fla. 32826
407-380-1553 or
800-345-2737
(2, 5)

**Law-Tech/Martial Arts Supplies Co.**
10711 Venice Blvd.
Los Angeles, Calif. 90034
213-870-9866
(10)

**Lifesaving Resources**
3087 PS Business Center
Woodbridge, Va. 22192
703-878-4503 or
800-221-3041
(20)

**Lift-All Co.**
102 S. Heintzelman St.
Manheim, Pa. 17545-1030
717-665-6821 or
800-433-0396
(17)

**LMB Hand Rehab Products Inc.**
P.O. Box 1181
San Luis Obispo, Calif. 93406
805-541-3992 or
800-541-3922
(9a)

**Long Island Productions**
925 Kearney St.
El Cerrito, Calif. 94530
510-528-1681
(1, 2, 3, 6, 7, 8, 9, 10, 11, 12, 13, 14, 15, 16, 17, 18, 19, 20, 21)

**Lyons Safety**
P.O. Box 1010
Germantown, Wis. 53022
800-558-8900
(1, 5a, 9a, 11, 12, 13, 14, 21)

**Magic Sorb/ITW Devcon**
30 Endicott St.
Danvers, Mass. 01923
508-777-1100
(9)

**Mansdorf & Associates**
4575 Hudson Drive
Stow, Ohio 44224-1773
216-686-5100
(6, 9, 13, 14, 16, 17, 18, 19)

**Marcom Group**
#4 Denny Road
Wilmington, Del. 19809
302-764-3400 or
800-654-2448
(1, 6, 7, 8, 10, 11, 12, 13, 14, 16, 17, 18, 19, 21)

**Marlin Industrial Division**
100 Kenna Drive
North Haven, Conn. 06473
203-239-5801
(7, 19)

**Media Group**
200 Taos Drive NE
Ada, Mich. 49301
616-956-9583 or
800-253-9239
(7, 19)

**Medical Devices International**
3849 Swanson Court
Gurnee, Ill. 60031
708-336-6611 or
800-323-9035
(3, 20)

**National Audiovisual Center**
8700 Edgeworth Drive
Capitol Heights, Md. 20743
301-763-1896 or
800-638-1300
(2, 10, 13, 19)

**National Safety Council**
1121 Spring Lake Drive
Itasca, Ill. 60143
800-621-7619
(1, 2, 3, 5, 6, 7, 8, 9, 9a, 10, 11, 12, 13, 14, 15, 16, 17, 18, 19, 20, 21)

**Neotronics of North America**
2144 Hilton Drive
Gainesville, Ga. 30501
404-535-0600 or
800-535-0606
(20)

**NETAC**
320 William Pitt Way
Pittsburgh, Pa. 15238
412-826-5320
(9)

**Network Inc.**
6649 Peachtree Industrial Blvd. #J
Norcross, Ga. 30092
404-441-1912 or
800-241-5689
(6, 7, 11, 14, 19)

**North American Crane Bureau**
1011 Executive Center Drive #218
Orlando, Fla. 32803-3520
407-896-2927 or
800-654-5640
(17)

**Occupational Medicine of York**
1601 2nd Ave.
York, Pa. 17403
717-843-5500
(12)

**Occupational Safety Training**
1400 N. Sportsmans Point
Inverness, Fla. 32650
904-344-4320 or
800-842-4142
(2)

**OMC Environmental**
4451-G Parliament Place
Lanham, Md. 20706
301-306-0632 or
800-662-7618
(21)

**Parallel Services**
7670 Pinesprings E. Drive
Indianapolis, Ind. 46256
317-577-9686
(2, 9, 13, 16, 19, 21)

**Personal Health Designs**
16 Timberland Drive
Loudonville, N.Y. 12211
518-458-8998
(9a, 14, 19)

**Positive Promotions**
222 Ashland Place
Brooklyn, N.Y. 11217
718-858-4199 or
800-635-2666
(1, 3, 7, 10, 11)

**Prinzing Enterprises**
30W270 Butterfield Road #120
Warrenville, Ill. 60555
708-393-2080 or
800-292-2914
(1, 5a, 13, 21)

**Recon Systems**
5 Johnson Drive
P.O. Box 130
Raritan, N.J. 08869-0130
908-526-1000
(9, 13, 21)

**Reflectory**
87 Broadway
Newburgh, N.Y. 12550
914-565-2037 or
800-654-1417
(7)

**Roco Corp.**
8254 One Calsis Ave. #240
Baton Rouge, La. 70809-3450
504-769-8877 or
800-647-7626
(5a, 20)

**Roytech Publications**
7910 Woodmont Ave.
Bethesda, Md. 20814
301-654-4281
(21)

**Safetec of America**
1333 Strad Ave.
N. Tonawanda, N.Y. 14120
716-692-9302 or
800-456-7077
(3)

**Safety & Health Alliance**
12021 W. Bluemound Road #120
Milwaukee, Wis. 53226
414-521-3323
(1)

**National Safety Council**
**Safety & Security Systems**
P.O. Box 247
Claymont, Del. 19703
302-798-5299
(6, 7)

**Safety Information Cinema**
5750 Wilshire Blvd. #530 W
Los Angeles, Calif. 90036
213-939-1181
(1, 2, 6, 7, 8, 9, 11, 16, 19, 20)

**Safety Kleen**
777 Big Timber Road
Elgin, Ill. 60123
708-697-8460
(9, 21)

**Safety Short Production**
2960 N. 23rd St.
La Porte, Texas 77571-3182
713-470-9999 or
800-458-2236
(2)

**Safety Software**
2030 Spottswood Road Suite 200
Charlottesville, Va. 22903-1245
804-296-8789 or
800-932-9457
(21)

**Safety Training Dynamics**
45 Stearns Ave.
Brockton, Mass. 02402
508-587-4984
(21)

**Safetycare Inc.**
4040 MacArthur Blvd. #300
Newport Beach, Calif. 92660
714-252-8005
(2, 3, 6, 7, 12, 16, 17, 18, 19, 21)

**Saunders Group**
7750 W. 78th St.
Minneapolis, Minn. 55439
612-944-1656 or
800-654-8357
(9a, 14)

**Savant Audiovisuals**
P.O. Box 3670
Fullerton, Calif. 92634
714-870-7880 or
800-472-8268
(13)

**Scantek**
916 Gist Ave.
Silver Spring, Md. 20910
301-495-7738
(18)

**Search Research Center**
7670 Pinesprings East Drive
Indianapolis, Ind. 46256
317-577-9686
(2)

**Seton Name Plate**
P.O. Box 1331
New Haven, Conn. 06505
203-488-8059 or
800-243-6624
(3, 7, 8, 9, 11, 14, 15, 16, 17, 18, 19, 20, 21)

**Sheridan Safety**
P.O. Box 2137
Dayton, Ohio 45401
513-233-4655 or
800-762-9330
(8, 9a, 12, 14, 15, 19)

**Sherwood Medical**
1915 Olive St.
St. Louis, Mo. 63103
314-241-5700 or
800-325-7472
(1)

**Simulaids**
12 Dixon Ave.
P.O. Box 807
Woodstock, N.Y. 12498
914-679-2475 or
800-431-4310
(20)

**Smith System Driver Improvement Institute**
1106 Pioneer Pkwy.
Arlington, Texas 76013
817-275-1225 or
800-777-7648
(7)

**Soltys Associates**
1741 N. Ivar Ave. #106
Los Angeles, Calif. 90028
213-465-5605
(11, 19)

**Soma-Sandler Occupational Medicine Associates**
275 Broadhollow Road #302
Melville, N.Y. 11747
516-756-2204
(6, 12, 13, 14)

**Southworth Products Corp.**
P.O. Box 1380
Portland, Maine 04104
207-772-0130 or
800-743-1000
(14)

**Spence-Geiger Associates**
1658 Cole Blvd. #260
Golden, Colo. 80401
303-233-8577
(21)

**START Rescue**
P.O. Box 1574
Southampton, Pa. 19866
215-357-8637
(3, 20)

**Summit Training Source**
620 Three Mile Road NW
Grand Rapids, Mich. 49504
616-784-4500 or
800-842-0466
(1, 2, 6, 7, 8, 9, 10, 11, 12, 13, 14, 15, 16, 17, 18, 19, 21)

**Talus Resources**
9245 Reseda Blvd. #740
Northridge, Calif. 91324
818-701-0364
(15)

**Tel-A-Train**
309 N. Market St.
Chattanooga, Tenn. 37405
800-251-6018

**Top Tape & Label**
3325 N. Service Road
Burlington, Ontario L7N 3G2
800-263-0580
(21)

**Trinity Corp.**
190 G Penrod Court
Glen Burnie, Md. 21061
410-766-5750
(11)

**Triple S**
141 Middlesex Turnpike
Burlington, Mass. 01803
617-273-2020 or
800-323-2251
(21)

**UES**
4401 Dayton-Xenia Road
Dayton, Ohio 45432
513-426-6900
(9a, 14, 15, 16)

**United Health Resources**
Cape Road
Hollis, Maine 04042
207-929-5100
(4, 8, 19)

**Valley Videos**
P.O. Box 379
Marne, Mich. 4935-0379
800-448-5889
(1, 19)

**Vasquez Management Consultants**
38760 Northwoods Drive
Wadsworth, Ill. 60083
708-249-1900 or
800-367-7378
(2)

**Visucom Management Systems**
P.O. Box 5472
Redwood City, Calif. 94063
415-364-5566
(2, 7, 14, 17, 19, 21)

**Weston, Roy F.**
1 Weston Way
West Chester, Pa. 19380-1499
215-430-7211
(17)

**Wiley, John & Sons**
**Interscience Division**
605 Third Ave.
New York, N.Y. 10158-0012
212-850-6000
(13)

**Windsor Group**
243 Old Stage Road
Spotswood, N.J. 08884
908-723-0366
(14, 19, 21)

**Work Stations**
165 Front Stm.
Chicopee, Mass. 01013
413-598-8394
(19)

**World Safety Organization**
305 E. Market St.
P.O. Box 518
Warrensburg, Mo. 64093
816-747-3132
(1, 14, 21)

**Writers Inc.**
1501 Casho Mill Road #1
Newark, Del. 19711
302-738-6226
(2)

**Yoder, KD, Associates**
23265 C R 28
Elkhart, Ind. 46517
219-875-7542
(13)

# INDEX